Anton Fuchs · Eugenius Nijman
Hans-Herwig Priebsch
Editors

Automotive NVH Technology

 Springer

Editors
Anton Fuchs
Area NVH & Friction
VIRTUAL VEHICLE Research Center
Graz
Austria

Hans-Herwig Priebsch
VIRTUAL VEHICLE Research Center
Graz
Austria

Eugenius Nijman
VIRTUAL VEHICLE Research Center
Graz
Austria

ISSN 2191-530X   ISSN 2191-5318   (electronic)
SpringerBriefs in Applied Sciences and Technology
Automotive Engineering: Simulation and Validation Methods
ISBN 978-3-319-24053-4   ISBN 978-3-319-24055-8   (eBook)
DOI 10.1007/978-3-319-24055-8

Library of Congress Control Number: 2015952984

Springer Cham Heidelberg New York Dordrecht London

Springer International Publishing AG Switzerland is part of Springer Science+Business Media
(www.springer.com)

SpringerBriefs in Applied Sciences and Technology

Automotive Engineering: Simulation and Validation Methods

Series editors

Anton Fuchs, Graz, Austria
Hermann Steffan, Graz, Austria
Jost Bernasch, Graz, Austria
Daniel Watzenig, Graz, Austria

More information about this series at http://www.springer.com/series/11667

Preface

The constant pressure towards the realization of environmentally friendly "green" vehicles is dramatically changing vehicle concepts and architectures. Examples are the growing importance of lightweight materials, the downsizing of internal combustion engines, and the introduction of more advanced after-treatment and combustion strategies including the use of alternative fuels and control strategies like "idle stop." We are witnessing continuous evolutions in automotive transmission technology characterized by a steadily increasing number of gear ratios combined with reduction of friction losses and dual clutch technology. And of course there is the "revolution" of alternative propulsion systems including pure electric (BEV) applications, gasoline or diesel-based hybrids (HEV), and plug-in hybrids (PHEV), each of them featuring many possible architectures and introducing new components like electric motors, power electronics modules, batteries, regenerative braking technology, integrated starter generator technology and additional gearsets.

This new scenario poses overwhelming challenges to the Noise, Vibration, Harshness (NVH) engineers, who are faced with systems of exponentially growing complexity exhibiting new multidisciplinary features with huge impact on NVH performances. Still, apart from being price-wise competitive, these new cars must also be appealing and meet the customers' taste and expectations with respect to comfort, fun to drive, and brand sound signature. There is consequently a growing demand for more advanced numerical prediction tools which can further support the NVH engineers during the concept and development phases of new vehicle design projects. Along with these powerful numerical tools, experimental analysis and diagnostics remain of paramount importance for a successful vehicle optimization.

The topics within the field of vehicle NVH are so numerous and diverse that it is practically impossible to give a complete survey within the framework of this book. In the next chapters a few selected topics will consequently be presented, which, although far from exhaustive, give a flavor of the diversity and complexity of the problems and solutions which characterize the field of vehicle NVH.

This book is structured into seven chapters, each of them authored by NVH experts from automotive industry and academia. These seven chapters are composed to give profound insights into many current vehicle development topics in the field of NVH. They focus on aeroacoustics, acoustics of geared systems, noise characterization and reduction of downsized engines, noise of electrified powertrains, lightweight exhaust systems and a substructuring method allowing to account for the dynamic interaction between the car body structure, the poroelastic trim material, and the interior cavity.

The topics addressed in the book also reflect the predominant trends in vehicle development towards reducing the number of hardware prototypes and "front-loading" by means of numerical simulation and enhanced physical component testing.

The majority of chapter contributions for this book are based on scientific papers presented at the 8th International Styrian Noise, Vibration, Harshness Congress (ISNVH 2014) in Graz, Austria and have been extended and refined for this book recently. The ISNVH congress itself is organized by VIRTUAL VEHICLE Research Center in cooperation with SAE International and supported by Magna Steyr and AVL List. The next ISNVH congress will be held from June 22 to 24, 2016 in Graz.

All paper contributions for the ISNVH congress and many more technical papers in the field of automotive NVH are published with SAE International and are available online from SAE International.

The editors gratefully acknowledge the support of SAE International and all chapter authors, which allowed to compile this book.

Graz, Austria   Anton Fuchs
July 2015   Eugenius Nijman
                                                                          Hans-Herwig Priebsch

About the Organization Editing this Book

VIRTUAL VEHICLE is an internationally operating research center that develops technologies for affordable, safe and environmentally-friendly vehicles for road and rail. With currently 200 employees, the center provides advanced know-how in a comprehensive range of technologies in vehicle development. The research activities cover specific single topics as well as complete vehicle systems. Since its foundation in 2002, VIRTUAL VEHICLE has established itself as a bridge between industry and science and as a well-respected research partner all over Europe and beyond.

VIRTUAL VEHICLE mainly focuses on developing new technologies, methods, and tools for the design and optimization of vehicles (automotive and rail). The vast majority of the R&D activities include a holistic system view, which helps avoiding very specific, isolated, and narrowly targeted developments that are widespread in today's vehicle research landscape. By considering rising complex requirements and their interactions with ever-increasing demands, VIRTUAL VEHICLE wants to guarantee for future-oriented solutions. Therefore, one main goal of the research center is providing in-depth know-how in single CAE disciplines and integrating this to a multi-domain design and optimization.

The NVH & Friction department of VIRTUAL VEHICLE deals with the fields of noise, vibration, and harshness (NVH), and aims to determine and reduce the friction loss in the vehicle. Owing to increasing system complexity, the research department seeks to handle the growing demands for drive systems with reduced fuel consumption, emissions, and weights, all while maintaining (or even improving) passenger comfort.

The major research topics of the NVH & Friction department are NVH material and technology, vehicle noise reduction, friction loss and vibration reduction, as well as flow acoustics—both by means of numerical simulation and testing and measurement. In combination with the other aspects of the full vehicle, these issues form an essential keystone in the development of future vehicle concepts.

Acknowledgments

The editors acknowledge the financial support of the COMET K2—Competence Centres for Excellent Technologies Programme of the Austrian Federal Ministry for Transport, Innovation and Technology (BMVIT), the Austrian Federal Ministry of Science, Research and Economy (BMWFW), the Austrian Research Promotion Agency (FFG), the Province of Styria, and the Styrian Business Promotion Agency (SFG).

Contents

Editors and Contributors

About the Editors

Anton Fuchs received the Dipl. Eng. degree in Telematics from Graz University of Technology in 2001 and Ph.D. degree in Technical Science in 2006 from Graz University of Technology. From 2002 to 2008, he was at the Institute of Electrical Measurement and Signal Processing, Graz University of Technology. Since 2008, Anton Fuchs has been associated with the VIRTUAL VEHICLE Research Center in Graz, Austria. He was Head of Thermo and Fluid Dynamics Department and coordinated the center development. In 2009, he received the venia docendi for "Process Instrumentation and Sensor Technology" from Graz University of Technology and became Associate Professor and Distinguished Lecturer. From 2012, he is Head of research department "NVH and Friction" at VIRTUAL VEHICLE Research Center. His main research interests include automotive sensing technologies. Anton Fuchs is author and coauthor of more than 100 scientific papers and patents. email: anton.fuchs@v2c2.at

Eugenius Nijman received his M.Sc. in Mechanical Engineering from Technical University of Eindhoven in The Netherlands in 1987. In the same year he joined "Low Noise Design Department" of TNO in Delft. From 1995, Eugenius Nijman was at Centro Ricerche Fiat (CRF), the research center of the FIAT Automobiles Group in Orbassano (Torino), Italy. In 1998, he became Head of the Engine NVH department at CRF and held this position till 2009. During that time he was responsible and especially involved in engine NVH optimization, NVH target setting and deployment, intake and exhaust noise, and impact noise phenomena. In 2009, he started his own consultancy company "Nijman Acoustics & Vibrations," that is based in Costigliole d'Asti (Italy). He also joined VIRTUAL VEHICLE Research Center in Graz (Austria) in 2011 and took over the position of the Scientific Head of the NVH department.

Eugenius Nijman is Head of the Scientific Programme Committee of the ISNVH Congress, chairman at different sessions at international acoustic conferences, reviewer for ISNVH and SAE, and author or coauthor of more than 40 scientific publications and patents. email: Eugene.nijman@v2c2.at

Hans-Herwig Priebsch received a Ph.D. degree from Graz University of Technology, Austria, in 1980. His research activities have been concerned with tribology and vibro-acoustics of power units. From 1981 to 1999, he worked for AVL List GmbH, Graz, Austria, where he was Manager in the business area of Advanced Simulation Technologies. Together with his team he developed the software AVL Excite aiming to simulate the vibro-acoustic behavior of engine and powertrain by means of elastic multi-body dynamics and nonlinear contact models. He is holding a Habilitation Postdoctoral qualification and has been giving lectures on machine dynamics and vehicle acoustics at Graz University of Technology, since 1994.

In 1999, he joined the Acoustic Competence Center, ACC Graz, where he has been responsible for all R&D work. Hans Priebsch has initiated and organized the International Styrian Noise, Vibration, Harshness Congress and has served as Chairman and Head of Scientific Programme Committee for this event. In 2008, ACC has been integrated into VIRTUAL VEHICLE Research Center as department "NVH and Friction." Hans Priebsch was the Scientific Head of this department until he retired in 2012. He is Scientific Consultant for VIRTUAL VEHICLE Research Center. email: hans-herwig.priebsch@v2c2.at

Contributors

Kwin Abram was introduced to the field of acoustics in 1989 while working as a student on gas turbine engines at Allison Engine Company. He now holds 20 patents related to the NVH field. He obtained a Master's degree in Mechanical Engineering from Purdue University in 1994 where he conducted diesel engine diagnostics at Herrick Laboratory and then worked for Cummins Inc. till 2001 conducting acoustics research. Kwin worked for 11 years on exhaust acoustics research for Faurecia Emissions Control Technologies and has recently returned to Cummins Inc. as an NVH Technical Advisor for diesel engines. email: kwin.abram@gmail.com

Christopher Albert studied physics and received his M.Sc. degree from Graz University of Technology in 2012. From 2011 to 2014, he did research on methods of acoustical measurements at VIRTUAL VEHICLE. He is now pursuing his Ph.D. in Theoretical Plasma Physics at Graz University of Technology. email: albert@tugraz.at

Dennis Bönnen studied vehicle engineering and received his Diploma degree from the University of Applied Sciences Cologne in 2003. He received his Ph.D. by conducting research on the investigation of vibrational behavior of brake discs from

Loughborough University in 2008. From 2007 he has worked in various NVH related positions at Faurecia Emissions Control Technologies. Currently, as Métier Manager, he is responsible for the methods and tools development within the predevelopment. email: dennis.boennen@faurecia.com

Stephan Brandl studied technical mathematics and received his M.Sc. degree from Graz University of Technology in 2004. He joined Acoustic Competence Center in 2004 and received his Ph.D. in the field of acoustic optimization by means of statistical energy analysis from Graz University of Technology in 2007. In 2008 he joined AVL and is now Product Manager for vehicle and powertrain NVH. email: stephan.brandl@avl.com

Alexandre Carbonelli graduated with Master's degree in Material Sciences from INSA de Lyon. He obtained his Ph.D. in Mechanical Engineering from École Centrale de Lyon and the Tribology and System Dynamics Laboratory (LTDS) with his research on vibro-acoustic phenomena of geared systems. This study has been performed in collaboration with Vibratec and Renault Trucks. He works now for Vibratec as a Research Engineer and is involved in vibro-acoustic simulations and research projects. email: alexandre.carbonelli@vibratec.fr

Rik W. De Doncker received his Ph.D. degree in Electrical Engineering from the Katholieke Universiteit Leuven, Leuven, Belgium, in 1986. In 1987, he was appointed as a Visiting Associate Professor at the University of Wisconsin, Madison. After a short stay as an Adjunct Researcher with Interuniversity Microelectronics Centre, Leuven, he joined, in 1989, the Corporate Research and Development Center, General Electric Company, Schenectady, NY. In 1994, he joined Silicon Power Corporation, a former division of General Electric Inc., as the Vice President of Technology. In 1996, he became a Professor at RWTH Aachen University, Aachen, Germany, where he currently leads the Institute for Power Electronics and Electrical Drives. Since 2006, he has been the Director of the E.ON Energy Research Center, RWTH Aachen University. email: post@isea.rwth-aachen.de

Christoph Fankhauser studied mechanical engineering and received his M.Sc. degree from the Graz University of Technology in 1994. He joined AVL in 1994, where he performed engine structural simulations and software coding for the thermodynamic gas exchange software BOOST. In 1999 he moved to the NVH department of MAGNA STEYR. After NVH advanced development and a group leader position in the comfort development, he transferred to the advanced development within MAGNA STEYR and performed several internal and external research projects, and among them several projects with the VIRTUAL VEHICLE. Since 2011, he has been the Head of MAGNA STEYR's NVH department. email: christoph.fankhauser@magna.com

Matthias Frank studied electrical engineering and audio engineering at University of Technology and University of Music and Performing Arts in Graz. After receiving his Diploma in 2009, he joined the Institute of Electronic Music and

Acoustics in Graz. In 2013, he finished his Ph.D. dealing with the perception of sound fields created by surrounding loudspeaker arrays. Matthias is member of the Audio Engineering Society and the German Acoustical Society. email: m.frank@kug.ac.at

Gottfried Grabner studied audio engineering and received his M.Sc. degree from the Graz University of Technology in 2003. He joined the MAGNA STEYR's department for NVH & Driving Comfort as acoustic specialist in 2003. From 2007, he continues to be Specialist in aeroacoustics development at MAGNA STEYR Engineering in Graz. email: gottfried.grabner@magna.com

Andreas Hofmann received the Diploma degree in Electrical Engineering from the Technical University of Munich (TUM), Germany, in 2010. Ever since, he has worked as a Research Associate at the Institute of Power Electronics and Electrical Drives (ISEA) at the Technical University of Aachen (RWTH), Germany. His main research interest is mitigating the acoustic noise of automotive traction drives particularly by control. email: post@isea.rwth-aachen.de

Barry James received his Master's degree in Manufacturing Engineering at Clare College, Cambridge, in 1992. Having started his career at Ricardo plc, he joined Romax Technology Ltd. in 1995 and has held a range of different positions, including Engineering Manager and Chief Engineer. In 2012 he was made Chief Technical Officer and is also Head of R&D. He has a strong interest in developing electro-mechanical drivelines for high power density, high efficiency, and low noise. email: barry.james@romaxtech.com

Joerg Jany studied mechanical engineering and received his M.Sc. degree from the Technical University of Munich in 2004. He joined the MAGNA STEYR's department for NVH & Driving Comfort as Acoustic Specialist in 2005. Since 2008 he has been team leader for vehicle acoustics. email: joerg.jany@magna.com

Matteo Kirchner received his B.Sc. degree in Industrial Engineering and his M.Sc. degree in Mechatronics Engineering from the University of Trento (Italy) in 2008 and 2011, respectively. From 2013 he is a doctoral student in vibro-acoustics at KU Leuven (Belgium) within the FP7 Marie Curie EID Project "eLiQuiD" (GA 316422). He spent the first 18 months of his Ph.D. track at VIRTUAL VEHICLE (Austria). email: matteo.kirchner@kuleuven.be

Christian Kranzler studied electrical engineering and audio engineering at Graz University of Technology and the University of Music and Performing Arts Graz (KUG) and recieved his M.Sc. degree in 2008. Christian worked as Researcher at the Institute of Electronic Music and Acoustics of KUG from 2009 to 2013. He joined the Acoustics Department of AVL List GmbH in 2013, and his topics are active sound generation and noise reduction in vehicles. email: christian.kranzler@avl.com

Gregor Müller studied mathematics at the Graz University of Technology with special focus on numerical mathematics and optimisation. He joined

MAGNA STEYR in 2005 as Simulation Specialist in Statistical Energy Analysis (SEA) and Wave Based Technique (WBT). In 2009 he received his Ph.D. for his work on modeling algorithms for the WBT. From 2011 he is also concerned with advanced development in the field of aeroacoustics. email: gregor. j.mueller@gmail.com

Joël Perret-Liaudet graduated in engineering and holds a Ph.D. from École Centrale de Lyon. He is Associate Professor in the Tribology and System Dynamics Laboratory at École Centrale de Lyon where he holds this position since 1992. He teaches courses in the fields of Mechanical Engineering and Solid Mechanics. In 1999, Joël Perret-Liaudet obtained an accreditation to supervise research in the field of dynamics and vibrations of parametric and/or nonlinear mechanical systems. He is interested mainly in research concerning the vibro-acoustic mechanisms, including the gear transmission, in its global and local approaches. Furthermore, his work focuses on understanding various dynamic, vibration, and acoustic phenomena coupling with contact mechanics and friction. email: joel.perret-liaudet@ ec-lyon.fr

Jan Rejlek studied electrical engineering and received his M.Sc. degree from Czech Technical University Prague in 2004. He joined VIRTUAL VEHICLE (former ACC) in 2004. He received his Ph.D. working on acoustic simulation by means of WBT from Graz University of Technology in 2012. Since 2014 he has been group leader of the "Vehicle Noise Reduction" group at VIRTUAL VEHICLE. email: jan.rejlek@v2c2.at

Emmanuel Rigaud graduated with Master's degree in Mechanical Engineering from INSA de Lyon. He became Research Engineer at École Centrale de Lyon and the Tribology and System Dynamics Laboratory (LTDS). In this context, he obtained a Ph.D. in Mechanics in 1998, and then became Assistant Professor in 2001. He holds an accreditation to supervise research since 2013. He is currently responsible for teaching in Mechanical Engineering at École Centrale de Lyon. He performs his research at the LTDS, within the Tribology, Physical Chemistry and Dynamics of Interfaces research team. His activities are centered on the relations between contact dynamics and the overall performance of mechanisms in terms of vibro-acoustic, friction, and wear. email: emmanuel.rigaud@ec-lyon.fr

Alois Sontacchi studied electrical engineering and audio engineering and received his Master's degree in 1999 from Graz University of Technology (TU Graz). In 2003 he obtained a Ph.D. degree in Technical Science at the TU Graz. He joined the Institute of Electronic Music and Acoustics at the University of Music and Performing Arts in Graz in 1999 and has been its Director since 2010. His research interests include psycho-acoustical perceptual evaluation, music information retrieval, as well as spatial audio signal processing and, the creation of virtual audio environment. email: alois.sontacchi@kug.ac.at

Hannes Steinkilberg studied mechanical engineering and received his degree (Dipl.-Ing. Univ.) at the Technical University Munich. In 2001, he joined Zeuna

Stärker, which later became part of Faurecia Emissions Control Technologies. Since then, he has worked in the field of Computational Fluid Dynamics and was team leader for NVH simulation. Currently, as Innovation Program Manager, he is responsible for Faurecia's Exhaust Dynamic Sound Technologies (Sound Generation and Noise Cancellation). email: hannes.steinkilberg@faurecia.com

Giorgio Veronesi graduated in Civil Engineering from the University of Ferrara in 2011. From 2012 to 2015 he was Ph.D. fellow of the Marie Curie Action project "Gresimo" at the VIRTUAL VEHICLE. He received his Ph.D. for a thesis on vibro-acoustic and porous material characterisation in 2015 from the University of Ferrara. From 2015, he is senior researcher in the "Vehicle Noise Reduction" group at VIRTUAL VEHICLE. email: giorgio.veronesi@v2c2.at

Michael Wiesenegger studied audio engineering and received his M.Sc. degree from Graz University of Technology in 2008. He joined the MAGNA STEYR's department for NVH & Driving Comfort as Acoustic Specialist in 2010. email: michael.wiesenegger@magnapowertrain.com

Franz Zotter received a Diploma in Electrical Engineering and Audio Engineering at University of Technology and University of Music and Performing Arts in Graz and joined the Institute of Electronic Music and Acoustics (IEM) in 2004. In 2009 he received a doctoral degree from the University of Music and Performing Arts in Graz. For his work on capture, analysis, and reproduction of sound radiation he was awarded the Lothar Cremer medal of the German acoustical society (DEGA) in 2012. Franz Zotter is the current Chair of DEGA's work group on virtual acoustics and co-director of IEM. email: f.zotter@kug.ac.at

Chapter 1
Assessment of the Vehicle's Interior Wind Noise Due to Measurement of Exterior Flow Quantities in Early Project Phases

Gregor Müller, Gottfried Grabner, Michael Wiesenegger, Joerg Jany and Christoph Fankhauser

Abstract The optimal styling of the exterior surface of a vehicle and its suspension system has a direct impact on interior wind noise. Both are determined in early project phases when typically no hardware prototype is available. Turbulent flows produce both external pressure fluctuations at the vehicle shell, known as hydrodynamic excitation, and sound waves, known as acoustic excitation. Hydrodynamic and acoustic sound sources are evaluated separately and relative to each other in the frequency domain in order to perform evaluations of different body shapes. The technical aim of the presented work is to investigate how acoustic quantities measured either directly in the exterior flow or as characteristic values of surface subsystems at the outside of a vehicle can be used to assess the influence of styling modifications to interior sound pressure level. The methodology is required to be capable of being integrated into the serial development process and therefore be quickly applicable. MAGNA STEYR Engineering has conducted extensive research to develop a method to ensure the best option is selected in early project stages.

Keywords Interior wind noise · Exterior styling modification assessment · Early project phase

G. Müller (✉) · G. Grabner · M. Wiesenegger · J. Jany · C. Fankhauser
Department NVH & Driving Comfort, MAGNA STEYR Engineering, Graz, Austria
e-mail: gregor.j.mueller@gmail.com

G. Grabner
e-mail: gottfried.grabner@magna.com

M. Wiesenegger
e-mail: michael.wiesenegger@magnapowertrain.com

J. Jany
e-mail: joerg.jany@magna.com

C. Fankhauser
e-mail: christoph.fankhauser@magna.com

© The Author(s) 2016
A. Fuchs et al. (eds.), *Automotive NVH Technology*,
Automotive Engineering: Simulation and Validation Methods,
DOI 10.1007/978-3-319-24055-8_1

1

1.1 Introduction

Interior noise targets for passenger car development programs are usually defined on a high level. Thus the entire chain of noise generation and propagation mechanisms has to be controlled in order to meet the requirements. At low vehicle speeds, rolling and engine noise needs to be carefully balanced, while at high speeds [1], low wind noise levels are critical in order to meet high acoustic quality standards. The implementation of robust sealing, insulation and damping in a concept design serves as a good basis for achieving these standards. Consequently, an acoustically tight vehicle becomes necessary to achieve a low wind noise contribution. This means that any unwanted airborne sound paths—vehicle acclimatization/ventilation paths excluded—must be closed.

The complete topic of exterior flow phenomena producing interior noise is subdivided into:

- low frequency excitation of pressure fluctuations exciting the vehicle's exterior surfaces and their structural resonances
- mid to high frequency excitations of exterior cavities by spoilers, struts, edges, especially at the underbody
- pressure fluctuations at sealings and side windows creating high frequency noise
- high frequency excitations by exterior flow sources or sources at exterior parts

This article is intended for the latter two phenomena.

Due to their physical properties the windscreen and side-window glasses are sensitive to sound transmission above 2 kHz and are main contributors to the interior noise. The NVH-wise optimal design of the exterior surface of the vehicle and its suspension system is an additional challenge. The exterior vehicle shape has a direct impact on the generated sound power as well as on the degree of turbulence along the external surface of the vehicle. However, exterior styling is usually determined in very early project phases, when no representative hardware of the vehicle is available. Also, the cost aspect has to be taken into account here. While additional sealing parts/lines or late modifications of the sealing system always lead to additional costs, simply changing the exterior shape of a vehicle is very often "complimentary" and therefore smarter. So the task is to assess exterior styling modifications with respect to interior sound pressure level. While usually this cannot be measured in styling finding phases, computational aeroacoustic (CAA) simulation robustly calculates the exterior, non-compressible fluctuations ("hydrodynamic part"). However, the calculation of these fluctuations along with the acoustic part, which is very often done as postprocessing step (e.g. Ffowcs-Williams Hawkings) is still costly, not to mention the coupling of the excitation to the structure.

Also many semi-empirical models such as the Corcos-model are still in use, aiming at the combination of measured or easy-to-simulate quantities. Until now however, no "overall" solution with reasonable time exposure and result quality could be found. Therefore the representation of aeroacoustic optimization as a pure measurement-based process is still a relevant issue.

Clay models, which physically represent a vehicle's exterior surface, are used to assess the exterior shape with respect to styling and aerodynamics. The models are used during wind tunnel sessions to assess important aerodynamic quantities such as the drag coefficient. The question arises, how these clay models can also be used for aeroacoustics. Of course, interior sound cannot be measured, so only measurement data gathered at the outside of the vehicle can be used.

1.2 Methodology

A passenger vehicle represents a large obstacle in the wind, so the exterior flow around the car gets turbulent and creates noise. In order to assess the wind noise generated sound pressure level in a passenger cabin, the different noise generation and transmission mechanisms throughout the frequency range of interest have to be properly considered.

Flow in a turbulent boundary layer causes pressure fluctuations generated by eddies which travel down the stream with convection velocity, known as 'hydrodynamic excitation'. These pressure fluctuations act as forces on vehicle structures and make them therefore vibrate. The same turbulent flow also radiates noise, known as 'acoustic excitation', and these sound waves also excite nearby plates. The situation gets even more complex, when additional parts like outer mirrors are mounted or when the flow must pass the a-pillar, where flow detachment might occur, followed by reattachment somewhere across the side window. In this case, additional hydrodynamic pressure fluctuations as well as broad band and a potential whistling noise of attachment parts contribute to the overall noise environment. Therefore the question arises, which of these wind noise mechanisms are the dominant ones for interior noise. Based on this evaluation, a measurement procedure can be defined to acquire a relevant set of quantities for interior sound pressure assessment.

For this purpose, the influence of hydrodynamics and acoustics is analyzed with the help of styling modifications. Wind tunnel tests as well as [2, 3] show that the difference of surface pressure fluctuation levels on side windows due to a shape modification in general does not coincide very well with interior sound pressure levels. Therefore, also the sound radiated in the near field of the vehicle is measured and analyzed in order to identify correlations with interior sound in a more general way.

The investigations were performed in the aeroacoustic wind tunnel at FKFS (Forschungsinstitut für Kraftfahrwesen und Fahrzeugmotoren, Stuttgart; engl.: Research Institute of Automotive Engineering and Vehicle Engines Stuttgart). Two test vehicles with different exterior shapes were selected to get a comprehensive overview among the different vehicle shapes: a Sports Coupè with a flat A-pillar and frameless door and a Sports Utility Vehicle with a steep A-pillar. The following measurement hardware was used:

Fig. 1.1 Ten surface microphones placed on the side window with a concentration on the hot spots of the side window fluctuations

(1) **Surface microphones** (10 resp. 20 mics, 1/2 inch, B & K, type 4949) on the left side window to measure the surface pressure fluctuations

The measurements were done using a different number of surface microphones. At the Sports Coupé 10 microphones were placed in regions with the highest pressure levels. This way the excitation mechanisms could be captured best (see Fig. 1.1). The focus was laid on

(a) the impact zone of the A-pillar vortex,
(b) the mirror wake, mirror stalk and mirror foot and
(c) the remaining area on the side window.

CAA simulation results showing the spatial pressure fluctuation levels (see Fig. 1.2) served as basis for the microphone positioning.

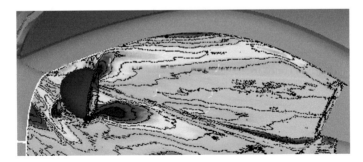

Fig. 1.2 CAA simulation (Exa Powerflow) showing pressure fluctuations on the left side window for a wind speed of 140 kph and 0° yaw angle. The A-pillar vortex reattachment zone and the mirror wake can be seen very clearly

Fig. 1.3 20 surface
microphones almost equally
distributed on the side
window

The SUV's side window was equipped with 20 microphones. This allowed a
more regular distribution than concentrating them on certain hot spots (Fig. 1.3).

All microphone measurement data were area averaged to get a single evaluation
curve for the whole side window excitation. Earlier measurements with a serial
vehicle confirm the assumption that the side window sound transmission is not
significantly influenced by the surface microphones. The result is shown in Fig. 1.4.

(2) **Concave mirror array** (108 mics) for evaluation of the radiated sound

This proprietary hardware from FKFS, described more in detail in [4], is suitable
for sound source localization. The mirror is located appr. 4 m sideways apart from
the driver's door outside of the flow and uses the principle of geometrical acoustics.

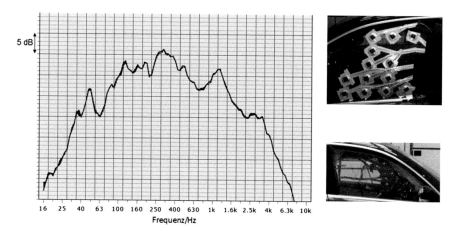

Fig. 1.4 Influence of surface microphones (right upper part) on interior sound pressure level
(driver's outer ear position). *Solid line*: with surface microphones, *dotted line*: w/o surface
microphones. The influence is negligible

Fig. 1.5 Sound radiation in the 3150 Hz third-octave band, measured near the driver's door outside the flow, using a concave mirror microphone array

The sound beams are focused by the mirror in the array and recorded using 108 microphones. Consequently no data post-processing like beamforming is required.

In Fig. 1.5 can be seen that the radiated sound field is dominated by the vehicle outer mirror and the A-pillar. In order to be able to assess a variant also in a quantitative way and not just qualitatively, one rating curve was derived by averaging the 108 channels to evaluate the styling modifications, which can be seen later on.

Of course the relevant sound direction for passengers is the inward one, while the concave mirror detects outgoing sound waves as well as sound waves reflected by the side panels of the vehicle. Because of the nonsymmetrical flow situation at a vehicle's side with respect to the flow direction as well as asymmetric attaching parts such as the outer mirror, one might expect a more or less strong directional sensitivity of sound radiation. In order to detect inward propagating waves, a larger number of microphones can be mounted flush on the side window. The acoustic part of the pressure fluctuations can be extracted by means of wavenumber decomposition [5]. The higher wave speed or longer wave length of acoustic waves compared to the hydrodynamic pressure fluctuations is used to get this information. However, this procedure needs a considerable amount of preparation effort (plate with microphones adapted to the side glass styling) as well as postprocessing effort. Furthermore, there are indications such as measurement results presented in [6] that this directional sensitivity was not strongly distinct.

One of the main intentions of the work presented in this paper is to define a testing methodology which focuses on quick analysis during testing.

(3) **Accelerometers** (6 resp. 5 sensors, uniaxial) on the side window

The sensors were uniformly distributed across the side window, 6 on the Sport Coupè's side window, 5 on the SUV's side window. The main purpose is the assessment of the direct excitation by hydrodynamics and acoustic waves from the outside. All measurement positions were averaged to get a single assessment curve.

(4) **Interior microphone** (1 microphone, driver's outer ear position).

1.3 Interpretation of Results

In the course of these investigations numerous exterior modifications were tested. Below, the two variants (1) 'larger mirror cap rear edge' of the Sport Coupè and (2) 'small obstacles on A-pillar' of the SUV are described more in detail.

The evaluation graph as in Figs. 1.7 and 1.9 consists of four pairs of curves (one pair for each measured quantity). Each pair shows the base and the variant data in dB over 12th octave bands. The interior sound pressure level as customer relevant criterion is A-weighted. All other quantities are linear. The question is to what extent the trend of the two indicator variables "concave mirror SPL" and "surface microphone's SPL" can be used to predict the two target variables 'acceleration level on side window' and 'interior noise SPL'.

The frequency range of interest starts above 1000 Hz where the wavelengths of the airborne pressure and the structure of the glass are starting to approach the coincidence. Below this range usually the sound transmission through other subsystems like underbody and door panels dominates and no dominance of the side window sound transmission is observed.

1.3.1 Larger Mirror Cap Rear Edge

Compared to the baseline design the air channel between the mirror cap and the mirror stalk tapers off more strongly towards the rear end (see Fig. 1.6). Figure 1.7 shows how the indicator variables and the two target variables behave relative to each other.

Fig. 1.6 Extension of the side mirror housing near the rear edge

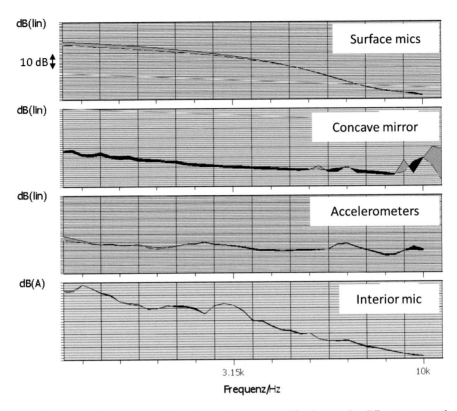

Fig. 1.7 The graph shows the effect of the mirror cap modification on the different measured parameters at 140 kph from ∼1 to 10 kHz (*light grey* = level decrease; *dark grey* = level increase). Indicating parameters: surface microphones and concave mirror. Target parameters: interior noise level and acceleration level, measured on the side window

The results for the mirror modification show a slight level increase in the range of $1 \div 2$ dB of the interior microphone while in some single frequency bands a small decrease is observed. The surface microphones, on the contrary, show a clear decrease of 2 dB for frequencies below 3.15 kHz while being unchanged for frequencies above. The concave mirror shows a constant increase of $1 \div 2$ dB for almost the complete frequency range under consideration. Only at very high frequencies ($9 \div 10$ kHz) major changes can be seen. The peaks directly originate from the outer mirror which was confirmed by additional measurements. The shift of the peaks in frequency is due to the change of size of the mirror cap. Measurements of different speeds showed that the peak is linearly related to the wind speed. By using a typical Strouhal number $Sh = 0.2$, the characteristic length L is ∼1 mm (given by $L = \frac{shv}{f}$, where v is the wind speed (more precise: the local free stream velocity) and f the frequency under consideration (in this case 9–10 kHz)). So, these peaks are probably caused by some geometric detail of the outer mirror.

As can be observed, the side window vibration correlates very well with interior sound. Summing up, none of the 2 indicator quantities are perfectly in line with interior sound. However the concave mirror shows a better correlation, as the slight level increase is shown in a very similar manner in the concave mirror SPL and interior sound SPL.

1.3.2   Small Obstacles on A-Pillar

Vehicles with a steep A-pillar tend to have a strong A-pillar vortex. The air flow within the vortex itself and the reattachment on the A-pillar causes high fluctuation levels, both acoustic and hydrodynamic. Therefore modifications of the A-pillar are promising in order to assess fluctuation level changes. This was done by application of small obstacles, so-called turbulators, on the upper A-pillar surface, see Fig. 1.8, which should not generate too much self-generated noise.

The basic idea is to change the flow detachment behavior of the A-pillar vortex and therefore alter the vortex itself. Substantial level changes of the two indicator variables were expected. In Fig. 1.9 the effect of these changes are presented.

The surface pressure decreases with 2–3.5 dB while all other curves show an increase, more or less constant over frequency. The clear decrease of surface pressures suggests that the structure of the A-pillar vortex has been changed substantially. The interior sound level increases, which correlates to the concave mirror. At this point it is not entirely clear whether this comes from the different shape of the A-pillar vortex or directly from the turbulator on the A-pillar. However, for frequencies above 4 kHz, the side window vibration level shows an even higher increase than interior sound, indicating that this path may well be not the only dominating one for the high frequencies.

Table 1.1 gives an overview of all conducted variants in the side window region.

Fig. 1.8 Modification of A-pillar shape: hemispheres of different sizes uniformly distributed on the upper A-pillar

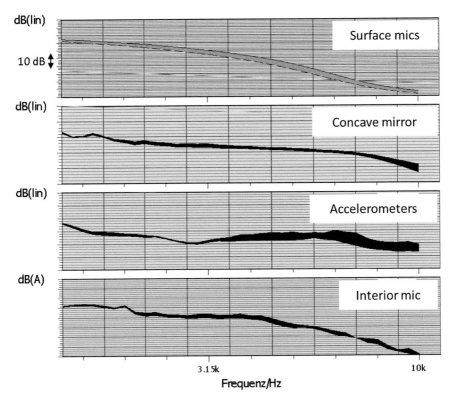

Fig. 1.9 Effect of the A-pillar turbulators on the different measurement devices at 140 kph from ~1 to 10 kHz (*light grey* = level decrease; *dark grey* = level increase). Indicating parameters: surface microphones and concave mirror. Target parameters: interior noise level and acceleration, measured on the side window

Summarizing, Table 1.1 reveals that the vibration level of the side window has an excellent correlation with interior sound. In 5 out of 7 cases, both the concave mirror array and the surface microphones predict the interior sound level change in a somehow satisfying manner. In 3 out of 7 cases, they contradict each other.

Clearly some simplifications were necessary to get this elementary description of a rather complex situation. First of all, just the frequency range from 1 kHz upwards has been taken into account. Furthermore, level changes of less than 1 dB were not considered. A positive correlation (a '+' in the table) means that the result of the corresponding measurement device shows the correct 'trend' for the assessment of the interior sound level change.

The following conclusions can be drawn:

1. The source of wind noise cannot be fully described by only one of the examined exterior flow indicator quantities.

Table 1.1 Overview over conducted variants with 2 vehicles

		Surface microphones	Concave mirror array	Accelerometers
	Styling modification			
Sport Coupè	Spoiler lip A-pillar	+	−	+
	Larger mirror cap rear edge	−	−	+
	No outer mirror	+	+	+
	Wheelhouse closed	+	+	+
	Wheelhouse closed, no outer mirror	+	+	+
SUV	Small obstacles on A-pillar	−	+	+
	Spoiler lip A-pillar	+	+	+
	Correlation with interior sound	**5/7**	**5/7**	**7/7**

A '+' indicates that the corresponding measurement device recorded the same change trend as the interior microphones, while a '−' indicates an opposite trend

2. Side window vibrations were found to be in good correlation with interior sound. As for now, it is proposed to use this quantity for interior sound level assessment, if interior sound cannot be measured directly (e.g. at clay models).

In order to further improve styling modification assessment, the following problems should be addressed in future:

In general the two exterior flow indicator variables show different behaviors. This suggests that the turbulences that produce the dominating radiated sound are not directly the ones on the side window surface. Those turbulences are probably located in the area of the air flow surrounding the mirror and in the volume of the A-pillar vortex itself. Therefore, the locations of the fluctuation microphones have to be rethought.

1.4 Summary

The presented evaluation methodology of hydrodynamic and acoustic sound sources was developed based on aeroacoustic R&D investigations. The relevance of correct interpretation of the measurement quantities for a successful exterior design has been clearly shown. The optimal aeroacoustic product design of individual components or complete vehicles can only be done efficiently when the complete aeroacoustic causal loop is actively managed. Based on these findings an efficient methodology can be defined, which enables the use of clay models for aeroacoustic purposes such as styling theme selection or the evaluation of exterior design modifications.

Acknowledgement The authors would like to thank Mrs. Eva-Elisabeth Spitzer for her excellent data preparation and analysis as part of her diploma thesis as well as Dr. Anton Falkner, Head of Integrated Vehicle Validation, and Johannes Mayr, Head of Virtual Development, for the trust placed in the aeroacoustic R&D activities.

References

1. Zeller P (2009) Handbuch Fahrzeugakustik. Vieweg + Teubner, GWV Fachverlage GmbH, Wiesbaden
2. Hartmann M, Ocker J, Lemke T, Mutzke A, Schwarz V, Tokuno H, Toppinga R, Unterlechner P, Wickern G (2012) Wind noise caused by the A-pillar and the side mirror flow of a generic vehicle model. In: 18th AIAA/CEAS aeroacoustics conference. American Institute of Aeronautics and Astronautics
3. Peng GC (2011) Measurement of exterior surface pressures and interior cabin noise in response to vehicle form changes. In: SAE international (2011-01-1618)
4. Helfer M, Bathelt H, Scheinhardt M, Sell H, Sottek R, Guidati S (2010) Messung und Analyse. In: Sound engineering in automobilbereich, Springer Berlin, S. 339–426
5. Hartmann M, Tokuno H, Ocker J, Decker F, Blanchet D (2014) Windgeräusch eines generischen Fahrzeugmodells: Synchrone Nahfeld-Fernfeld und Fernfeld-Innenraum Messungen sowie Simulationen. DAGA 2014, Oldenburg, Deutschland
6. Ask J, Davidson L (2005) The near field acoustics of a generic side mirror based on an incompressible approach. research report, Department of Applied Mechanics, Chalmers University of Technology

Chapter 2
Sound Optimization for Downsized Engines

Alois Sontacchi, Matthias Frank, Franz Zotter, Christian Kranzler and Stephan Brandl

Abstract Today, the number of downsized engines with two or three cylinders is increasing due to an increase in fuel efficiency. However, downsized engines exhibit unbalanced interior sound in the range of their optimal engine speed, largely because of their dominant engine orders. In particular, the sound of two-cylinder engines yields half the perceived engine speed of an equivalent four-cylinder engine at the same engine speed. As a result when driving, the two-cylinder engine would be shifted to higher gears much later, diminishing the expected fuel savings. This chapter presents an active in-car sound generation system that makes a two-cylinder engine sound like the more familiar four-cylinder engine. This is done by active, load-dependent playback of signals extracted from the engine vibration through a shaker mounted on the firewall. A blind test with audio experts indicates a significant reduction of the engine speed when shifting to a higher gear. In the blind test, experts favored the interior sound of the proposed sound generation system and perceived better interaction with the vehicle.

Keywords Downsized engine · Active sound generation · Interaction

A. Sontacchi (✉) · M. Frank · F. Zotter
Institute of Electronic Music and Acoustics, University of Music
and Performing Arts Graz, Graz, Austria
e-mail: sontacchi@iem.at

M. Frank
e-mail: frank@iem.at

F. Zotter
e-mail: zotter@iem.at

C. Kranzler · S. Brandl
AVL List GmbH, Graz, Austria
e-mail: christian.kranzler@avl.com

S. Brandl
e-mail: stephan.brandl@avl.com

© The Author(s) 2016
A. Fuchs et al. (eds.), *Automotive NVH Technology*,
Automotive Engineering: Simulation and Validation Methods,
DOI 10.1007/978-3-319-24055-8_2

2.1 Introduction

Down-sized engines create a significantly different sound in the passenger cabin in
comparison with traditional four-cylinder (and greater) combustion engines [1–4].
The observed differences are mainly caused by the different characteristics of
engine orders due to engine structure and combustion sequence. In [5], basic
coherences between engine orders and elements of music are discussed. Sound
attributes and perceived timbre in regard to consonant and dissonant are treated
considering engine order intervals and related harmonics. In addition, it has to be
mentioned that the amplitude distribution of engine orders mainly determine the
perceived pitch of the overall sound [6].

In regard to reduce the fuel consumption the optimal theoretical gear change
should happen at around 2000 revolutions per minute (rpm). Studies under practical
conditions show for an examined two-cylinder engine (FIAT 500[1]) that the typical
gear change occur almost at 4000 rpm.

By comparing measured run-up spectra (cf. Fig. 2.1) of the examined (a) two-
and a typical (b) four-cylinder combustion engine under full-load condition we can
address the following: Differences of partly missing, reduced or enhanced engine
orders can be observed. In case of the two-cylinder engine, the 1st engine order is
dominating whereas the 2nd order is most emphasized in case of the four-cylinder
engine.

In order to obtain a resulting four-cylinder engine sound in the passenger cabin
of the two-cylinder vehicle, we have to acoustically introduce or eliminate several
defined engine orders. Subsequently, possible strategies are presented and their
feasibility under practical considerations is discussed. Out of these strategies, one
promising approach has been selected and implemented. The implementation yields
the desired magnitude profile of the most relevant engine orders.

Series of tests show that the average gear change is positively influenced. The
switching moment is moved towards lower rotational speeds, i.e. the gear change is
executed at lower rpm.

Moreover, subjective evaluation of the drivability considering dynamic handling
aspects exhibit improved results without modifications on the vehicle itself but
solely acoustically supporting a four-cylinder sound in the passenger cabin.

Conducted experiments partly took place at public transport routes under normal
road traffic conditions and at a special dedicated test track.

---

[1]The test car was a Fiat 500 0.9 TwinAir Turbo with 82hp built in 2009.

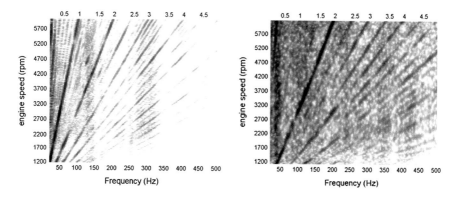

Fig. 2.1 Measured sound pressure spectra in passenger cabin. Spectra are normalized to the corresponding engine order maxima and depicted sound pressure dynamic is limited to 50 dB. Run-up under full-load condition (*left*) at the top, of examined two-cylinder gasoline engine (FIAT 500) and (*right*) at the bottom, an arbitrary four-cylinder gasoline engine (AUDI A3). At first glance, dominant 1st order in the upper and 2nd order in the lower case will be easily recognized

2.2   Strategies

2.2.1   *Additional Synthesized Engine Pulses*

A straight forward method would be the playback of a four-cylinder sound by the audio system of the car, whereby the played back sound has to mask above all the first order of the original engine sound. However, this means that the level of noise in the car interior increases. Best masking is achieved when every second pulse of the played back sound coincides temporally and locally with the original pulses, at least at the position of the driver. Besides the required complexity of the template related sound synthesis of the additional pulses in order to satisfy a natural-sounding result, both temporal and local coincidences within the vehicle would overexcite a practical solution (cf. [7] for an exhaustive discussion about theoretical limits).

2.2.2   *Active Noise Cancellation*

To get a controlled condition in the car interior, we want to cancel the engine sound or noise in the frequency areas, where we later want to add the relevant engine orders of a four-cylinder car. This means that we have to cancel the noise in the band between 20 and 80 Hz. Therefore, it is important to know, if we are basically able to cancel the frequencies in that area. For this reason we conducted sweep measurements in the passenger cabin on various positions (with a 10 cm spaced lattice microphone array and a dummy head) in the supposed movement area of the head of the driver and the co-driver.

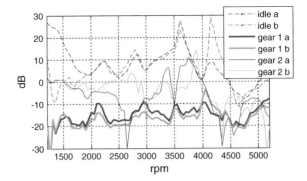

Fig. 2.2 Amplitude of the 1st engine order after noise cancellation. The mean amplitude- and phase relation of two trials (a and b) in the 1st gear is taken as ANC filter. For the two trials in the second gear and especially in the idle mode, there are 1st order enhancements, because the transfer paths from the engine to the car interior are not consistent over different gears and loads

A detailed description of these executed measurements, as well as documented measurement setup, procedure and discussion of the evaluated data in regard to Active Noise Cancellation (ANC), can be found in [7].

Examining the measured sound pressure level at the driver position for various run-ups, considering different gears and loads, attested a huge variability of the transfer function from the engine to the car interior (cf. Fig. 2.2). As a consequence, in order to perform a reliable ANC system, the coupling of the gearbox and the powertrain has to be considered in real time applications.

Moreover, different operating and occupation states will result in different transfer functions as well. If these states are not considered in the ANC coefficients, noise boost can occur, cf. Fig. 2.3 and [7]. Thus, successful ANC requires fast

Fig. 2.3 Maximum noise reduction with 1 or 4 passengers using theoretical ANC coefficients optimized for the condition with 1 passenger only (4 loudspeakers + shaker)

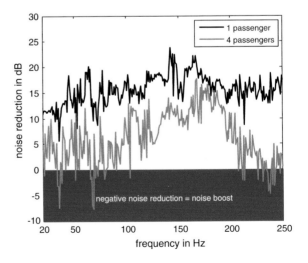

adaptation that is continuously informed by multiple distributed error microphones in the car interior and sensors capturing the engine's vibrations.

2.2.3 Engine Order Emphasis

In order to enrich the existing two-cylinder engine sound in the passenger cabin and result in a perceived four-cylinder engine sound the important engine orders (e.g. 2nd order of the two-cylinder engine sound) are emphasized. This is done by capturing the actual engine load condition, amplifying the relevant (missing) frequency bands, and playing them back to the passenger cabin.

To capture the operational-dependent properties of all relevant engine orders, proper sensor locations have to be determined.

In Fig. 2.4, the Campbell diagram of the most promising and finally selected sensor position is shown. The fully occupied spectrum provides a good chance to retrieve all required engine orders that might be missing in the observed car interior sound.

The playback of the generated sound should be perceived to be coming from the engine (active sound generation, ASG). Therefore, a shaker at the firewall and four loudspeakers of an ordinary in-car sound system were used. According to the carried out measurements described in [7], the shape of the sound field produced by the shaker is more balanced than one produced by the loudspeakers, albeit the shaker's frequency response is limited to low frequencies. As sound field control is

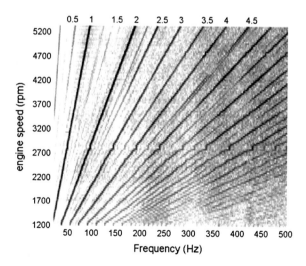

Fig. 2.4 Run-up spectrum under full-load condition of examined two-cylinder gasoline engine (FIAT 500). Measurement with accelerometer sensor mounted on a specific engine position. The resulting Campbell diagram exhibit distinct integral multiple engine orders, as well as half and partly quarter engine orders

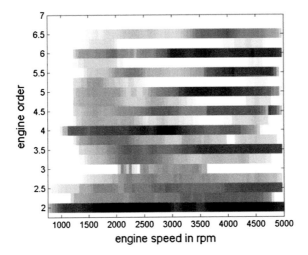

Fig. 2.5 Normalised amplification patterns of individual engine orders. Depicted dynamic range is limited to 50 dB, whereby black indicates 0 dB and white −50 dB or less. Enhancement of the partly missing engine orders is dependent of engine speed (rpm). Various load conditions are tackled via capturing at the specific sensor position mounted on the engine. Transfer functions of sensor and shaker are considered as well

in fact more interesting at those frequencies, our study suggests using shakers for active sound generation rather than loudspeakers.

In order to adjust the required amplification for each of the missing engine orders (cf. Fig. 2.5) the balance of the actual and target Campbell diagram has been determined. Furthermore, the transfer functions of the capturing sensor and reproducing shaker/loudspeakers have been measured and considered, too.

The amplification of the relevant missing engine orders can be controlled via a graphical interface running under the software Pure Data.[2]

The amplification of the engine orders also depends on the engine speed. Figure 2.5 shows the emphasis of the engine orders over the engine speed for the considered two-cylinder car (FIAT 500) to obtain the intended four-cylinder sound.

In Fig. 2.6 the principle implementation concept is sketched for one engine order. Missing engine orders are obtained from the above introduced acceleration sensor signal via dynamic band pass filters

The required extremely narrow-band filter characteristics are implemented with cascaded low- and high-pass filters. In order to prevent audible clicks (artefacts) during dynamic adjustments the subsequent sections are adapted at different time stamps (cf. indicated timing periods in (ms) in Fig. 2.6). Therefore, artefacts will

---

[2]Pure Data (Pd) is a visual programming language for creating interactive computer music and multimedia works. Pd is an open source project with a large developer base working on new extensions. It is released under a license similar to the BSD license. It runs on GNU/Linux, Mac OS X, iOS, Android and Windows.

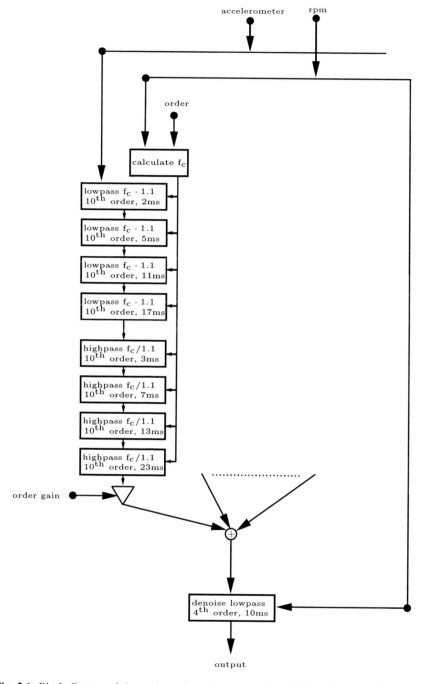

Fig. 2.6 Block diagram of the engine order enhancement. Four 10th order low- and high-pass filters extract the predefined orders from the accelerometer signal from the engine

result in a low-frequency band-limited noise that can be further reduced with a
subsequent low-pass filter (denoise-stage).

Amplification of the individual engine orders takes places as depicted in Fig. 2.5.
Dynamic adaptation towards engine speed is tackled with the rpm signal from an
inductive voltage transformer mounted on the drive line and measured entries in a
table look-up scheme.

Playback in the passenger cabin mainly employs with the shaker at the firewall
and the in-car sound system. As loudspeakers, the 4 loudspeakers of the Fiat's stock
in-car sound systems were used on an Alpine PMX F460 4-channel amplifier. The
shaker was a Sinuslive Buss-Pump II on a Raveland XCA-400 4-channel amplifier.

2.3 Resulting Sound

In Fig. 2.7, the measured spectra of a run-up under full-load condition are shown.
The upper spectrum (a) depicts the two-cylinder sound in the passenger cabin.
Spectrum (b), at the bottom, exhibits the resulting modified sound with active sound
generation in parallel.

A visual comparison of both spectra lets distinguish a significant enhancement of
the 2nd engine order at least within the important speed range from 1500 up to
3000 rpm.

Within this speed range the active sound generation causes an increased loudness level. However, the resulting loudness trend over engine speed (rpm) is much
more stable and balanced (cf. Fig. 2.8). In addition, the resulting loudness trend
over engine speed much better fits to the intended four-cylinder target sound.

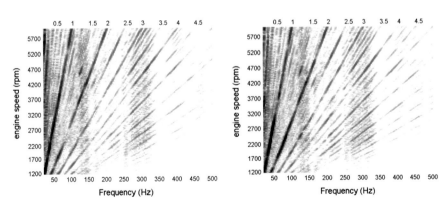

Fig. 2.7 Measured sound pressure spectra in passenger cabin. Spectra are normalized to the
corresponding engine order maxima and depicted sound pressure dynamic is limited to 50 dB.
Run-up under full-load condition (*left*) at the top, of examined two-cylinder gasoline engine (FIAT
500) without active sound generation and (*right*) at the bottom, resulting sound spectrum of
examined engine superimposed with active sound generation

Fig. 2.8 Measured loudness trend over engine speed for both real two-cylinder sound (*black curve*) and resulting sound with active sound generation mimicking the target four-cylinder sound (*gray curve*)

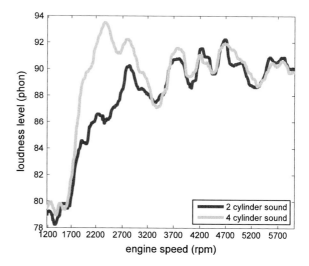

2.4 Experiment

2.4.1 Method

The implemented sound enhancement from the previous section has been evaluated by 10 subjects. The subjects were asked to drive the test car at two different dates. For both rides, the subjects drove the same two-cylinder car (FIAT 500), but for one of the two tours, the four-cylinder sound was played back. However, the subjects were not informed about the changes made to the car—solely active sound generation was activated or turned off. Conducted experiments partly took place at public transport routes under normal road traffic conditions and at a special dedicated test track. Although the subjects were asked to drive rounds on a test circuit, only the way to the test circuit is considered for the analysis of gear shifts in the following. This specific test design is caused by the fact that the driving behavior under common conditions shall be investigated.

Arrived at the test side, the subjects had to perform a reduced AVL standard drivability test procedure at the test circuit. Afterwards, the subjects were asked to evaluate the car in terms of drivability and acoustical aspects (cf. Fig. 2.14).

2.4.2 Test Track

The route is presented in Fig. 2.9 and has a length of 16.1 km from which 4.5 km are city roads and 11.6 km are outside the city limits. The vehicle speed over the driven distance is shown in the subsequent Fig. 2.9 for all subjects. From the CAN-bus of the car, the consumption, the vehicle- and the engine speed were recorded as depicted in Fig. 2.10.

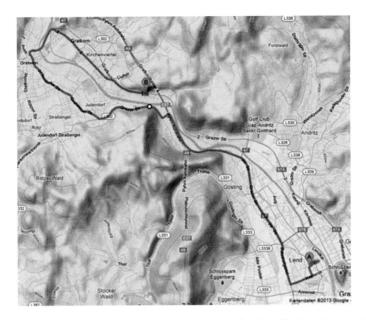

Fig. 2.9 The test track leads from AVL in Graz to the test circuit in Gratkorn. It has a length of 16.1 km [map from Google Maps: https://www.google.at/maps]

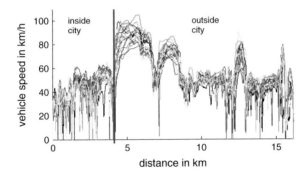

Fig. 2.10 Vehicle speed over driven distance. The first 4.5 km are driven within the city limits with a maximal speed of 60 km/h. The remaining distance is driven outside the city limits with maximal speed up to 100 km/h

Finally, after both tours, subjects were asked to fill in a general form about date of driving licence, approved driving licence categories, driving experience (amount of kilometres), driving performance per annum, ownership of a car (in case of affirmation: brand, driving performance p.a., typical driving profile: city, country side, motorway).

2.4.3 Engine Speed at Gear Shift

Besides a subjective evaluation of the test persons, the driving test investigated if the subjects shift into higher gears at lower rpm. Since no real time information about the gear is available, the gear shift event (in regard to the fuel consumption: only into higher gear) has to be detected over a discontinuity in the ratio between the engine speed and the vehicle velocity. Figure 2.11 shows this gear-shift ratio together with the normalized engine speed in rpm.

More interesting is the rpm-value at which the subjects shift into the higher gears. Therefore, we determine the maximum engine speed in the last 2.5 s before every gear-shift as sketched in Fig. 2.12.

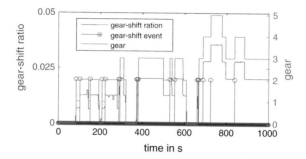

Fig. 2.11 The gear shift is detected over discontinuities in the gear-shift ratio, which is the ratio between the engine speed and vehicle velocity

Fig. 2.12 At every gear-shift event, the maximal engine speed in the last 2.5 s is determined

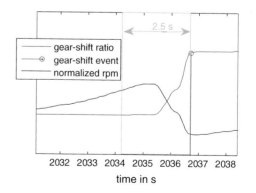

2.5 Results

For every subject and for every gear, these determined engine-speeds are compared between the test drives with the two- and the four-cylinder sound. In the fourth gear, there is a significant difference in the maximal engine speed between the two- and the four-cylinder sound. Figure 2.13 shows that the subjects in median shift at about 260 rpm earlier into the 4th gear when driving with the four-cylinder sound. For all other gears, no statistically significant improvement can be observed. This can be explained because the first three gears are mainly used in city traffic, where gear shifting depends on outer circumstances and the 5th gear has hardly been used on this test track. The shift into the 4th gear however mainly depends on personal decisions. The result of the test thus supports the hypothesis that the sound of the car is a major indication for the gear shift.

The same tendency can be observed when examining the average engine speed in each gear for individual subjects, cf. Fig. 2.14. The average speed in the 2nd and 4th gear decreases in the case of the four-cylinder sound. Moreover, subjects spent more time in the 4th gear.

After both test drives, the subjects were asked to evaluate the car in terms of drivability and acoustical aspects. The mean evaluation of all aspects is shown in Fig. 2.15. For most aspects, no significant difference can be observed. However, in all cases with significant differences, the car with the four-cylinder sound was perceived as better. Especially the car interior noise was evaluated as better with a p-value <0.01. It is notable that also the tip-in and tip-out are perceived as better when driving with the four-cylinder sound, although no changes have been made to the engine itself. This again emphasizes the importance of the engine sound for the driving experience and that a four-cylinder sound is perceived as having better quality than a two-cylinder sound.

Fig. 2.13 Engine-speed difference when subjects shifted to the fourth gear for the test car with a two- and a four-cylinder sound

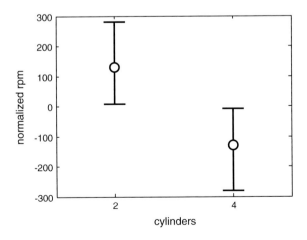

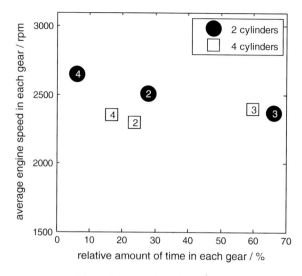

Fig. 2.14 Average engine speed in each gear and relative amount of time in each gear for the test car with a two- and a four-cylinder sound, exemplarily for one subject

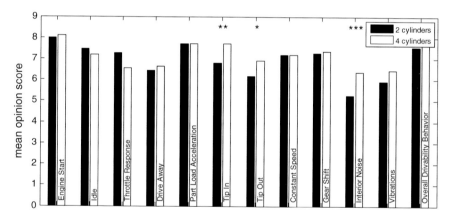

Fig. 2.15 Mean results of the questionnaire, asterisks indicate the significance level of a Mann–Whitney U-test: * $p < 0.1$, ** $p < 0.05$, ***$p < 0.01$

2.6   Conclusion

Three strategies were investigated to create a two-cylinder sound. In all strategies a shaker at the firewall is used as additional sound source. The first strategy is to play back synthesized combustion pulses, the second is to cancel the 1st engine order by destructive superposition and the third is to enhance engine orders which emphasises a four-cylinder sound.

For the first strategy, we recorded combustion pulses up to 3000 rpm in order to extract template pules for the synthesis playback. However, for a single full-load condition already 40 templates would be necessary to explain 90 % of the variations in the combustion pulses. Moreover, to achieve perfect masking the played back sound must fulfil temporally and locally constrains that can be hardly tackled at all passengers positions nor in various realistic conditions at the driver position itself.

For the second strategy, it is important that the engine as well as the shaker on the firewall create a homogeneous sound field in the car interior especially around the position of the driver. We therefore measured the phase relation between the shaker and specific positions in the car interior with a 24-channel microphone array in a dimension of 50×30 cm. The phase deviations in the measured area are small enough for the relevant frequencies (which are lower than 100 Hz). However, the amplitude and phase relation between engine vibrations and the interior noise is not consistent over different gears and loads. A mean amplitude and phase relation has to be chosen since information about the current gear in not available in real time. In some cases, this mean amplitude and phase relation leads to amplification instead of a cancellation of the 1st order.

The more promising strategy of the four-cylinder sound enhancement has been evaluated in a test scenario. Ten people drove a test track once with the original two-cylinder sound and on a different date with the four-cylinder sound enhancement without being informed about the changes made to the car. The subjects found that the tip-in and tip-out responded better in the car with the four-cylinder sound and they perceived the sound as better. The recorded rpm also show that the subjects shifted into the 4th gear at a lower rpm with the four-cylinder sound as. For the other gears, there are no significant differences between the rpm values because the first three gears are mainly used in city traffic, where gear shifting depends on outer circumstances and the 5th gear has hardly been used. The shift into the 4th gear however mainly depends on personal decisions. The result of the test thus supports the hypothesis that the sound of the car is a major cue for the gear shift. Furthermore, the proposed active sound generation system will also improve the driving comfort in case of automatic transmission with downsized engines.

Acknowledgement This work was supported by the project ASD, which is funded by Austrian ministries BMVIT, BMWFJ, the Styrian Business Promotion Agency (SFG), and the departments 3 and 14 of the Styrian Government. The Austrian Research Promotion Agency (FFG) conducted the funding under the Competence Centers for Excellent Technologies (COMET, K-Project), a program of the above-mentioned institutions.

References

1. Brandl S, Graf B (2011) Sound engineering for electric and hybrid vehicles. Procedures to create appropriate sound for electric and hybrid vehicles. In: 1st international electric vehicle technology conference, Yokohama
2. Graf B, Resch M, Maunz C, Dolinar A, et al. (2011) Sound engineering for downsized engines. In: JSAE annual spring congress, Yokohama
3. Graf B, Brandl S, Rust A (2012) Sound solutions for downsized powertrains. In: DAGA 2012, Darmstadt
4. Brandl S, Graf B, Rust A (2012) NVH challenges and solutions for vehicles with low CO2 emission. In: 7th international Styrian noise, Graz. doi:10.4271/2012-01-1532
5. Alt N, Jochum S, Sound-Design unter den Aspekten der Harmonielehre der Musik, MTZ 1/2003, Jahrgang 64. doi:10.1007/BF03226679
6. Terhardt E, Stoll G, Seewann M (1982) Algorithm for extraction of pitch and pitch salience from complex tonal signals. J Acoust Soc Am 71:679–688
7. Frank M, et al. (2014) Comprehensive array measurements of in-car sound field in magnitude and phase for active sound generation and noise control. SAE Int J Passeng Cars – Electron Electr Syst 7(2):596–602

Chapter 3
Reducing Noise in an Electric Vehicle Powertrain by Means of Numerical Simulation

Barry James, Andreas Hofmann and Rik W. De Doncker

Abstract The noise performance of fully electric vehicles is essential to ensure that they gain market acceptance. This can be a challenge for several reasons. Firstly, there is no masking from the internal combustion engine. Next, there is pressure to move to cost-efficient motor designs such as Switched Reluctance Motors, which have worse vibro-acoustic behaviour than their Permanent Magnet counterparts. Finally, power-dense, higher speed motors run closer fundamental frequency to the structural resonances of the system [1]. Experience has shown that this challenge is frequently not met. Reputable suppliers have designed and developed their "quiet" sub-systems to state of the art levels, only to discover that the assembled E-powertrain is unacceptably noisy. The paper describes the process and arising results for the noise simulation of the complete powertrain. The dynamic properties are efficiently modelled as a complete system and subjected to motor excitation (torque ripple, electro-magnetic forces and rotor imbalance). Innovation in this project comes from the speed of the modelling and analysis, so that analysis and data interpretation comes early enough in a project to be effective in reducing the noise problems. This contrasts with the approach of simulating problems that have already occurred in testing. Actions to reduce the motor noise are explained and identified. System dynamic response identifies the operating points in which different excitation mechanisms are most problematic and steps are taken to reduce the dynamic response. Also, problematic conditions can be identified where innovative motor control algorithms are necessary.

Keywords Switched reluctance motors · Noise reduction · Dynamic response

B. James (✉)
Romax Technology Ltd, Nottingham, United Kingdom
e-mail: barry.james@romaxtech.com

A. Hofmann · R.W. De Doncker
RWTH Aachen University, Aachen, Germany
e-mail: post@isea.rwth-aachen.de

R.W. De Doncker
e-mail: post@isea.rwth-aachen.de

© The Author(s) 2016
A. Fuchs et al. (eds.), *Automotive NVH Technology*,
Automotive Engineering: Simulation and Validation Methods,
DOI 10.1007/978-3-319-24055-8_3

3.1 Introduction

The long term targets for the automotive industry to reduce CO_2 emissions provide a substantial incentive to develop and manufacture electric and hybrid vehicles and to bring them to the mass market, as opposed to producing niche vehicles for low volume. For this to be achieved the new vehicle technologies need to appeal to the wider market and satisfy all the quality aspects in current products.

Over recent decades, passenger cars have continually improved their noise, vibration and harshness (NVH) performance through the development and implementation of a range of methods for product design, analysis, development and manufacture.

In real terms, the price of vehicles has continued to fall while quality and performance has increased. The price volatility and uncertainty of supply of rare earth material makes Permanent Magnet Synchronous Machines (PMSM's) less attractive as a potential solution for mass production.

This combination of quality and price sets a high target for electric and hybrid vehicles to match. In recent years many companies have looked to alternative motor technologies, such as Switched Reluctance Motors (SRM's) as a potential solution to mass producing motors cheaply. Running a motor at higher speeds allows greater power density, giving lower weight, smaller motor and lower overall cost.

However, SRM's have worse vibro-acoustic behaviour, in that their current waveforms are non-sinusoidal. This makes SRM's have very high harmonic content when comparing with standard rotating field machines. Also the radial forces which mainly cause stator vibration are rather strong as torque is produced by reluctance force and not Lorentz force. The harmonics are proportional to speed. High rotational speed thus makes low order harmonics hit the structural resonances earlier. Low order harmonics are usually quite strong, so that high speed provides an additional significant challenge to the design of a suitable product.

3.2 Common Approaches to Simulation

Methods for designing quiet products has progressed substantially over the past quarter century. The development of Computer Aided Engineering (CAE) tools such as Multi-Body Dynamics (MBD), Finite Element Analysis (FEA) and Multi-Domain Simulation tools have been developed. The intention is that the dynamics of systems can be simulated and problems can be identified and solved without having to manufacture hardware.

The excitation from the machine is mainly due to electro-magnetic forces. These forces are usually calculated analytically [1] or numerically [2–5]. The numerical approach is usually split into a software tool to discretely solve the differential equations which describe the electrical machine and which accounts for the geometry and the electromagnetic nonlinearities. The vibration response is that provided by a 3D structural FEA or MBD.

The state-of-the-art approach for vibro-acoustic simulation of the electrical machine is a coupled co-simulation between the tool which calculates electro-magnetic behaviour and the FE tool which determines the structural behaviour [6]. Hence, the structural FEA has to be solved in each time step. This makes the approach time consuming and thus hardly possible for a complete powertrain. Also full run-ups or even whole driving profiles are almost impossible. Using this model to guide the design process and thus eliminate problems before they occur cannot realistically be considered.

MBD packages and Multi-Domain Simulation tools are frequently used for time domain simulations [7–10]. Again, the problem exists that the modelling and analysis is too slow for the results to be useful in guiding the design process.

In its implementation in FEA, the natural frequencies and the dynamic response of a system to a given excitation can be calculated [11–13]. It is now generally accepted that to simulate phenomena whose excitation is inherently periodic (gear whine, torque ripple, imbalance) it is more efficient to simulate in the frequency domain [14]. The system is linearised and the loss of any non-linear behaviour is more than offset by the speed of analysis and the improvement in interpretability of the results. Transfer functions give an instantly interpretable indication of the behaviour of the system across a wide range of speed conditions.

When simulating for natural frequencies and dynamic response, the definition of the "system" often follows the division of responsibility within the design process. The motor manufacturer will focus on the motor system and simulate the rotor shaft, bearings, housing etc. The motor excitation of the torque ripple, radial forces and rotor imbalance will be included. Likewise the gearbox manufacturer will focus on the gearbox shafts, bearings, gears and housing, with the excitation coming from the gear transmission error [11–13].

There are several problems with using generalist tools such as FEA and MBD for the simulation of noise in complex systems. Firstly, modelling the system is time consuming. The use of automeshing has alleviated this but this is really only applicable to linear, homogeneous mechanical components such as shafts, and is of little use in the modelling of non-linear components such as splines, gear meshes and rolling elements bearings. These non-linear components indicate the second problem, in that such modelling requires high levels of expertise, which is in short supply and has a high associated cost in all corporations. Analysis times are long, and the data can be difficult to interpret, which means that the process of analysing the system, interpreting the results and iteratively searching for an optimised solution is slow.

The result is that such NVH analysis is, at best, only applied to problems that are identified during hardware testing and need resolving and not to the avoidance of problems during the design process. Thus the original purpose of CAE is not achieved.

The division of design responsibility can lead to problems. For example, torque ripple is not usually in itself an NVH problem so long as the "system" consists of the single shaft of a motor. However, once this is passed through the gear pairs this torsional excitation is converted into radial excitation, leading to housing vibration and radiated noise. Some papers [15] have shown how modelling the motor alone implies no problem with the motor noise, whereas including the full driveline indicates a noise propagation mechanism for the torque ripple through the gearbox housing. The risk is that this is only discovered once the prototype hardware is assembled.

3.3   Simulation Methodology Used

The introduced methodology works in the frequency domain. It calculates the eigenvectors and eigenvalues of the complete structural system of the powertrain. From this, various analyses are possible from which a range of engineering assessments can be made.

Transfer functions can be calculated between different points, giving rise to a fundamental understanding of how the system behaves and why the system responds in a particular way. The methodology makes use of the principle of modal superposition in calculating the system response. From this the resulting vibration due to a number of excitations and their harmonics can be calculated, and the operating deflected shape can be viewed, to give the resulting vibration at any frequency.

The key targets of this methodology are that (i) the speed of the process should be such that the results are available within a time frame that they are useful for making design decisions and (ii) the results can be interpreted easily so as to guide engineering design decisions [16].

With regard to the electric motor, the excitation comes in the form of force shapes that arise from the electromagnetic simulation. Details are given below (Fig. 3.1).

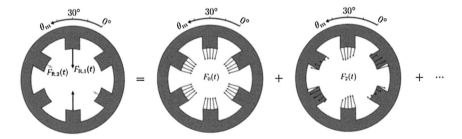

Fig. 3.1 The force shapes that are derived from the electro-magnetic simulation are expressed as a spatial Fourier series

3.3.1 Electromagnetic Motor Simulation

The main objective of the electromagnetic simulation is to provide the force-shape amplitudes for the modal superposition. Thus the model of the electrical machine has to calculate the relevant forces in the machine and spatially decompose them into the force shapes. These force shapes are the modal excitation of the structure's eigenmodes.

The forces which influence the acoustic behaviour of the machine are usually radial force. The introduced method, however, works also for tangential and axial forces. An SRM can be modelled as described in [2, 17]. The phase-voltage equation

$$u(t) = Ri(t) + \frac{d\psi(t)}{dt} \tag{3.1}$$

is integrated by using a multi-domain simulation tool to solve it for the phase-flux linkage ψ. R is the ohmic resistance of a phase and i is the applied current. The flux linkage is used together with the rotor position θ to calculate all kinds of forces such as radial force $F_R(\theta, \psi)$, torque $T(\theta, \psi)$ etc. Their relation with position and flux linkage is highly nonlinear but algebraic. Thus, it can be calculated in an electromagnetic FE preprocessing step and stored for online look-up in the simulation tool. This step incorporates the geometry of the SRM and the material properties of the used steel.

The time depending forces in the air gap (here, exemplary the radial force) can be expressed as a Fourier series upon the spatial position α:

$$F_R(\alpha, t) = \sum_{\nu=-\infty}^{\infty} \underline{F}_\nu(t) \cdot e^{j(\nu\alpha)} \tag{3.2}$$

$\underline{F}_\nu$ is the complex Fourier coefficient of the force shape with the order ν. Experience has shown that only a few of the lowest-order force shapes actually contribute considerably to the overall vibro-acoustic characteristics, so that the series in (3.2) can be truncated when $|\nu|$ is between 15 and 20 [4].

The resulting force shapes $\underline{F}_\nu$ vary with time. Thus, a temporal Fourier analysis is performed to determine the excitation $\underline{F}_\nu(\omega)$ of the respective mode with order ν. This excitation has to be weighted by the vibration response of the respective mode shape ν of the drive-train structure [4].

The eigenvectors of the motor present a challenge since the laminated stack and windings do not behave as isotropic materials that can be easily modeled. Some people have sought to deal with the anisotropic behavior of the stack [1], and the author has used this approach in the past in an attempt to match simulation and test data, but the result was unconvincing.

Others take the view that the principal behaviour of the motor that is being studied are vibration shapes in the 2D plain of the motor and these two space directions are isotropic.

In this instance, the full 3D motion is considered as the motor is part of the full drivetrain system. The method used is to simulate the stator as a solid body with reduced mass density and reduced Young's modulus. These are reduced by the so-called stacking factor which is the ratio between steel (normal electrical silicium steel) and insulation coating. This factor depends on the width of the steel sheets and in this case is about 0.96.

The coils can be modelled as solid copper bodies but with reduced mass density, as there is insulation material and most importantly air between the windings. It is difficult to claim that the copper bodies add any stiffness, so if they are to be included then they are done so only as an additional mass.

Recent work that compares test and simulation results on a SRM has shown that good correlation can be achieved even when the copper windings are ignored, hence this is the approach that was used here.

3.3.2 System Response

The force shapes are applied as excitations to the drivetrain structure and the dynamic response found using the principle of modal superposition.

Once the motor excitation has been calculated it can be applied to a dynamic model of the system to calculate the system response and hence surface vibration and radiated noise.

The frequency of the excitations vary. Maximum motor speed sees imbalance at 400 Hz (once per revolution) and Mode 6 Force Shape at 19200 Hz (48 cycles per revolution). It is true that all simulation methods struggle to maintain accuracy at high frequency levels and the author is well aware of these limitations from a wide range of practical projects. Discussions with other experts have led to the opinion that "we all face the same limitations in physics".

Nonetheless, the approach is still valid for the purposes for which it was intended. At this stage of the design the aim is not to produce the perfect simulation, rather to indicate the major problems, identify the best performing layout and provide guidance to the design team on vibration reduction. This approach to simulation is applied to a wide range of methods and have been embraced by industry experts who espouse "Failure Mode Avoidance" rather than perfect simulation.

Within the context of this design project it is better to understand the principal noise generation mechanisms and reduce them, rather than to proceed to prototyping in ignorance of what the vibration peaks may be.

It has been established that it is important to model the full system. In this case this means not just the motor but the gearbox components and housing, mounts, driveshafts and power electronics housing. This way an understanding of the full system response can be derived and it is possible to avoid the problems that have arisen from simulation of the separate sub-systems [15].

For the system modelling, a proprietary design and simulation package for drivelines is used to efficiently model the shafts and bearings of the gearbox and motor, the gears, mounts, driveshafts, stator and the housing of the gearbox, motor and power electronics.

Details of the modelling and analysis method have been covered in many previous publications [14, 16, 18, 19]. Like many FEA based approaches, the method involves multiple, coupled shaft systems, with a linear model used to derive mode shapes that are forced in the frequency domain. The housing is efficiently included by taking an FEA model of the housing and performing a dynamic condensation using Craig-Bampton modal reduction. Component Mode Synthesis is used to link the reduced dynamic model of the housing to the dynamic model of the internal components [20].

The key difference with methods based purely on generalist FEA packages is that specific algorithms are used to calculate the stiffness of the non-linear components such as rolling element bearings, gear meshes and splines, leading to solution times that are many times faster. Tailor made post-processing routines facilitate the interrogation of the analysis results.

Accuracy of this method has been proven in many projects over the years and has been proven for EV applications [4]. Implementation within OEM's has shown that the time for NVH modelling and simulation can be reduced by 80 % compared to generalist FEA tools [18]. It is this speed of analysis and interpretation that is so important if NVH simulation is to assist the design process (Fig. 3.2).

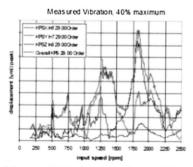
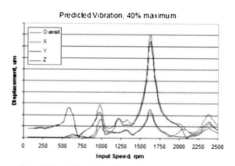

Measured vibration from the test data, 40% maximum drive torque

Predicted vibration from RomaxDesigner®, 40% maximum drive torque

Fig. 3.2 Comparison of test and simulation results for vibration for an EV powertrain [4]

3.4 Candidate Project for Practical Implementation

The key innovation described in this paper is not that noise of the powertrain can be simulated, but that it can be simulated in a timescale that allows performance improvements to be achieved within the timescales and constraints of an active design project.

In this instance the design project is ODIN (Optimized electric Drivetrain by Integration), an FP7 funded project which is led by Bosch for the design of a highly integrated electric vehicle driveline that will be installed into the Renault Zoe. Details of the project can be found on the public portal [5].

At the outset of the project the key innovation targets were:

- Use of high speed SRM (up to 23000 rpm)
- High degree of integration between the motor and the gearbox
- Single lubrication/cooling system for the whole drivetrain
- High efficiency gearbox, optimised for NVH

The project deliberately set out to move into unknown territory with regard to motor speed etc. in order to create a design that would challenge what is conventionally accepted as standard practice. In doing so, it places a substantial requirement on designing for low noise, and with it the need to simulate, identify and minimise noise within the timescales of a standard design process.

3.5 Initial Simulation: Concept Design

The project set out to simulate the noise characteristics of the driveline at the very earliest opportunity. It was known that the gearbox arrangement would affect the propagation of motor noise to the outside, so in selecting the gearbox layout and even defining the bearing positions, dynamic simulations were carried out to understand the system behaviour.

At this stage unit excitations were used, so as to understand the system behaviour. This included torque ripple, radial forces and imbalance from the motor and transmission error from the gears. Representative values for each were collected from specialists amongst the project consortium, based on their past experience.

The representative unit excitation values were not the actual values, since the gears and motor were yet to be defined in sufficient detail for the actual excitation to be calculated. Nonetheless, they were sufficient since the project was not, at this stage, aiming to give an absolute prediction of the vibration. Rather, it was aiming to understand the system response, guide design improvement and compare one layout with another (Fig. 3.3).

The complexity of this process needs to be highlighted. At this stage of the design process there is no housing, so no value of surface vibration or radiated noise can be calculated. The approach used here was to calculate the total sound power transmitted through the bearings and use this as a metric in guiding the concept selection (Fig. 3.4).

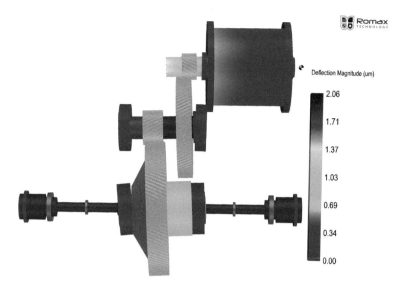

Fig. 3.3 An example of the initial dynamic simulation, to illustrate the level of model complexity used in selecting the concept for dynamic performance

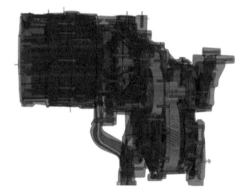

Fig. 3.4 The full model of the production E-Powertrain that was used to verify the method of simulating the dynamics of the internal components for concept design

The analysis team needed to check that this approach was valid, so an existing EV driveline (in production, complete with housing) was subjected to the same analysis method. A model of the internal components, equivalent in detail to the concept model in ODIN, was created and the same analysis applied. The results for the total vibratory power through the bearings for the simple model was compared to that for the original model, complete with fully detailed housing design (Fig. 3.5).

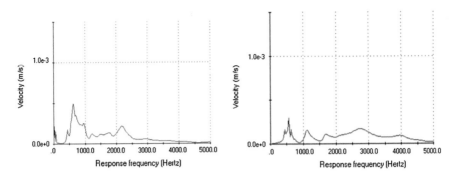

Fig. 3.5 Comparison of the system response to unit excitation for (i) the model of the internal components alone and (ii) complete system model (see Fig. 3.4) to verify that the simulation of the internal components was giving valid guidance

After careful development of this method it was confirmed that this simplified method was able to give results that were satisfactorily similar to those from the analysis of the fully detailed design.

3.6 Intermediate Simulation: Concept Selection

Once the concept layout for the rotating components had been selected, there was still the decision as to how the driveline and power electronics were to be assembled into the vehicle. Two options were identified, referred to as "T" and "L" layouts (Fig. 3.6).

Naturally, it was not possible to fully design a housing for both layouts; the project timing required that the choice had to be made without either design being fully modelled. Therefore, a simple, representative housing was modelled for both structures and the system simulation carried out.

This time the simulation was more involved. The driveline mount stiffnesses were included and the assessment was made based on total structure borne vibration, measured at the mounts, and by summing the housing kinetic energy, which is indicative of the total radiated noise. The same excitations were used as in the initial simulation (Fig. 3.7).

Fig. 3.6 Definition of "L" and "T" driveline layouts

Fig. 3.7 System model for the "T" Layout, showing simple housing model and mount locations

The simulation was carried out using representative values of excitation. It was possible to identify which concept had the best fundamental dynamic behaviour. For the selected concept, it was possible to compare which noise mechanism (torque ripple, radial forces etc.) was most significant at each speed and also identify the problematic modes of vibration, associated with the peaks. Figure 3.8 shows the response for the different excitation sources, each of which have different

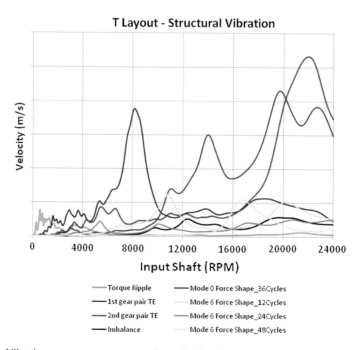

Fig. 3.8 Vibration response to representative unit vibrations across the full speed range, for all vibration sources

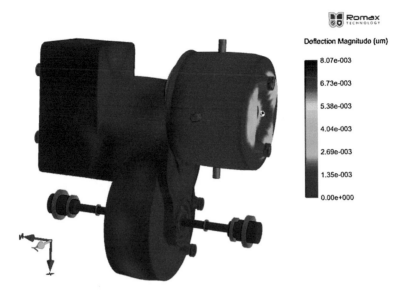

Fig. 3.9 Screen shot of the animation of the operating deflected shape (ODS) for the key vibration peak, Mode 0 Force Shape, 36 cycles per revolution

frequencies. The frequency excitation at maximum motor speed varies from 400 Hz for imbalance to 19.2 kHz for Mode 6 Force Shape, 48 cycles.

Feedback was given to the design team to change the housing design so as to minimise these modes of vibration.

This feedback is commonly given in the form of animated Operating Deflected Shapes (ODS) of the vibration. A screen shot of one such animation can be seen in Fig. 3.9. The complex vibratory motion of the end face of the motor housing, at the far right end of the model, was predicted by the simulation of the compete powertrain structure and it is known that this matches the experience of engineers who have worked on previous motor design projects. As well as providing feedback to the design process for the application of ribs in this area, this indicated that the system simulation was matching the test data from previous projects.

3.7 Detailed Simulation: Housing Design

The housing design was developed for the selected concept, with the structure modified and ribs applied to reflect the feedback from the intermediate simulation. The updated housing design was included and the system model and the reduction in vibration compared to the original design observed.

Figure 3.10 shows the vibration results for the Mode 0 Force Shape, 36 cycles, across the speed range. It shows that the levels of predicted vibration have

Fig. 3.10 Comparison of vibration due to mode 0 force shape, 36 cycles, between intermediate and detailed models, showing reduction achieved through targeting specific vibration behavior

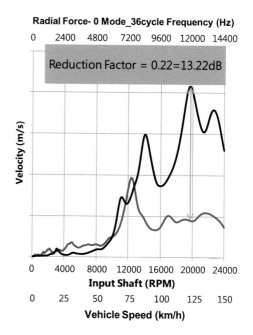

significantly reduced and thus the specific intention of the design-analysis-redesign iteration has been achieved.

It is true to say that some reduction would have occurred without such guidance from the intermediate simulation, since the concept housing has no ribs and the final housing does. However, it is better to design the housing with knowledge of guidance on likely noise problems rather than have to design in ignorance of where such problems may occur.

3.8 Further Simulation: Refinement

The final stage of using this simulation was in the refinement of the design. The vibration from all major sources, including harmonics, was be inspected and compared across the speed range.

Not all vibration results improved between the intermediate and detailed models. New peaks appeared and it was possible to inspect why these occurred.

Figure 3.11 highlights an area of the structure that is particularly active in response to Mode 6 Force Shape, 24 cycles. The main area of activity is where the oil tank is to be located for the dry sump system, an area that did not exist in the intermediate model. This issue was highlighted to the design team who were able to consider mitigation steps in this particular area for the final design.

Up to this point, the system response was calculated due to representative unit excitations, irrespective of speed and load. Now the detailed design of the

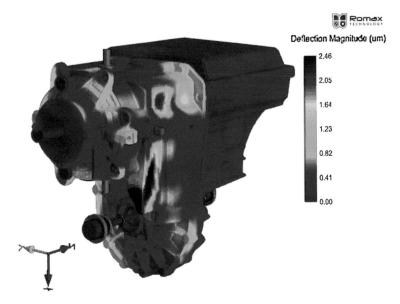

Fig. 3.11 Vibration due to mode 6 force shape, 24 cycles. An example of a new area of concern that has arisen in the transition to the detailed model, and which is being targeted through further design optimisation

components that cause the excitations (motor, gears) was complete, more accurate values for the excitation, dependent on speed and torque, could be applied.

Thus, feedback can be given to the design of the components that cause the excitation—the peaks due to gear transmission error could be identified and care could be taken to design the gear micro-geometry to minimise transmission error at these load conditions. Similarly for the motor, the problematic modes of excitation that existed were targeted and the control strategy was designed with the specific intention of eliminating these peaks.

Automotive-size SRMs usually suffer from two particular radial mode shapes. One has the same order as the number of poles within one phase, because the radial force is highly concentrated at the poles and discretely pulls the stator together there. The second shape is the so-called "breathing mode" with the order $\nu = 0$. This eigenmode is an exception. As its corresponding force shape is uniform upon the air-gap position α, this mode is excited by the superposition of all phase forces. This is very similar to the more familiar tangential excitation which we call "torque ripple". The two excitations are the same besides their direction.

The well-known Direct Torque Control [21] approach tries to produce smooth torque to overcome the effects which are cause by torque ripple. Direct Instantaneous Force Control does the same regarding overall radial force [22]. It controls the overall radial force smoothly and thus keeps force shape zero constant in time, i.e. $\underline{F}_0(t) = $ const. The design of the machine can be chosen so that mode 0 is the main noise source within the machine by choosing the number of pole pairs to

Table 3.1 Excitation reduction by DIFC compared to standard control

	21 Nm		110/50 Nm	
	36th cycle	72nd cycle	36th cycle	72nd cycle
5,000 rpm	−36 dB	−26 dB	−44 dB	−23 dB
8,000 rpm	−31 dB	−22 dB	−28 dB	−27 dB
12,500 rpm	−24 dB	−14 dB	−22 dB	−15 dB

be sufficiently high [30]. DIFC thus has the potential to eliminate this main noise source without any additional hardware expenses.

Hysteresis current control is quasi-standard to control SRM's. It produces a considerable ripple in the overall force signal. In comparison, DIFC promises to significantly reduce mode-0 borne noise. Table 3.1 shows the simulated reduction of force-shape 0 in a few operating points.

21 Nm is the most often used torque in the load cycle of the investigated machine. 110 Nm is the maximum overload torque. 5,000 rpm is a typical low-speed operating point and 8,000 rpm is the corner point. 12,500 rpm is where the 36th cycle hits the mode-0 resonance. The 36th and 72nd cycle are the most prominent mode-0 excitations for this kind of high-speed motor. 12,500 rpm is relatively high speed such that the machine can only produce 50 Nm in this operating point.

The simulation results in Table 3.1 assume perfect knowledge of machine parameters. They thus have to be understood as best-case approximation. However, parasitic effects as the switching of the inverter or current sensors have been taken into account. Even though parameter uncertainty degrades would degrade the benefit of DIFC a little, these results underline the potential of DIFC to basically eliminate mode-0 borne vibration.

Other mode shapes, however, stay fairly untouched such that other noise sources than mode 0 can be assumed to be equivalent to the case with standard control.

Nevertheless, the control strategy exhibits rather high torque ripple. While this may not be a problem in the machine, it can cause significant noise due to propagation through the structure of, say, the gearbox [15]. This issue can easily be investigated by means of the proposed vibro-acoustic simulation method. The control only has to be implemented in the electromagnetic motor simulation and the rest has to be performed as described above.

The graph shown in Fig. 3.8 shows that at low speed the torque ripple is represents the highest vibration source in terms of its contribution to structural vibration, but is less significant at higher frequencies. The insight from this system simulation is used to guide the development of the motor control and indicate what method should be applied to which speed range.

3.9 Initial Testing

At the time of writing some initial testing of the motor as a stand-alone unit had been carried out and some initial data is becoming available. This did not allow a full assessment of the methodology for simulating the system dynamic response since this would require the gearbox to be included. However, it was possible to compare the vibration for different control strategies.

At 21 Nm and 6000 rpm, an attenuation of −22 dB was achieved on 72nd cycle vibration by using DIFC compared to standard control. This compared to the original predicted attenuation of −26 dB at 5000 rpm.

At 21 Nm and 12000 rpm, an attenuation of −14 dB was achieved on 36th cycle vibration by using DIFC compared to standard control. This compared to the original predicted attenuation of −24 dB at 12500 rpm. It is believed that the reduction at higher speed is compromised a little since control quality suffers at the high speed than in the simulation.

3.10 Outcome

At the time of writing the paper, the full drivetrain had not been manufactured and so the success of this methodology for reducing the dynamic response has not been completely verified. Indeed, perhaps the success or otherwise of the method cannot be fully verified since it will never be known what the performance would have been without it.

However, it has been shown that at successive stages of the design process the predicted peak levels of vibration have been systematically targeted and reduced. In itself, this represents a development that goes beyond what has been previously achieved.

The full process can be represented with regard to the V-model for system design and development which is commonly used as a reference frame across industry. It can be seen that we are targetting insight at all points in the design process (the left hand part of the "V"), when it is cheaper and easier to fix, rather than wait until the bottom of the "V" or, worse, until the end of the development phase when the full powertrain is finally assembled (Fig. 3.12).

It is recognized that the most effective use of CAE is in the prediction and prevention of problems, yet previously the process of NVH simulation has been too slow to do this and it has been limited to correlation and elimination of problems (once they have already occurred). This is a significant change in approach and one that can significantly assist EV powertrains in achieving the level of design refinement required to make them acceptable to the mass market.

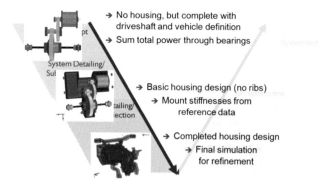

Fig. 3.12 Development and maturation of the simulation model with reference to the V-model systems approach for design and development

3.11 Summary

The development of CAE tools has often focused on achieving greater capability of simulation. The belief is that by modelling and analysing the world with greater levels of precision, greater insight will be achieved and CAE will achieve its goal of engineering better products.

However, the risk is that by targeting greater precision, the modelling and analysis process becomes too slow and the results too difficult to interpret. Engineering insight is lost and the results are not available in time to inform the design process.

This paper demonstrates that by using the correct analysis tools, and by creating the right model at the right time, it is possible to simulate the noise from motors and to systematically target noise reduction during the design process. This represents a significant change from the approaches used to date and directly targets the aim of CAE—the engineering of better products.

Acknowledgement The research work was carried out as part of the ODIN project (Optimized electric Drivetrain by Integration) and was co-funded by the Seventh Framework Programme of the EC.

References

1. Jordan H (1950) Geräuscharme elektromotoren, 1st edn. Girardet
2. De Doncker R, Pulle D, Veltman A (2011) Advanced electrical drives—analysis, modeling, control. Springer
3. Bösing M, Hofmann A, De Doncker RW (2014) Universal acoustic modeling framework for electrical drives. In: The 7th IET international conference on power electronics, machines and drives, 2014

4. Grunwald A James, B (2010) System approach to consider NVH, efficiency and durability in the optimisation of an electric all wheel drive gearbox. In: 9th international CTI symposium, December 2010 Berlin
5. http://www.fp7-odin.eu/consortium.htm
6. Van der Giet M, Schlensok C, Schmulling B, Hameyer K (2008) Comparison of 2-d and 3-d coupled electromagnetic and structure-dynamic simulation of electrical machines. IEEE Trans Magn 44(6):1594–1597
7. Garcia O, Kargar K, Renault, simulation tool for transmission and driveline systems design. SAE 2000-01-0832
8. Ciesla C, Jennings M, Ricardo North America. A modular approach to powetrain modelling and shift quality analysis. SAE 950419
9. Rivin E, Analysis and reduction of rattling in power transmission systems. SAE 2000-01-0032
10. Rust A et al (1990) Investigations into gear rattle phenomena—key parameters and their influence on gearbox noise. In: Proceedings, institution of mechanical engineers. C404/001
11. Donley M et al (1992) Dynamic analysis of automotive gearing systems. SAE 920762
12. Gradu M et al (1996) Planetary gears with improved vibrational behavior in automatic transmissions. VDI Berichte Nr. 1230
13. Lim T, Houser D (1997) Dynamic analysis of layshaft gears in automotive transmission. SAE 971964
14. James B, Douglas M (2002) Development of a gear whine model for the complete transmission system. 2002-01-0700
15. Meier C, Electric drive acoustics - a challenge for the mercedes B-class. In: International conference noise optimization EV/HEV 25.-27.09.2013
16. Platten M (2012) An engineer-led approach to driveline dynamic CAE. In: 7th international styrian noise, vibration & harshness congress, 13–15 June 2012, Graz, Austria
17. Carstensen C, Fuengwarodsakul NH, De Doncker R (2007) Flux linkage determination for correct modeling of switched reluctance machines - dynamic measurement versus static computation. In: IEEE international electric machines drives conference, 2007, IEMDC '07. 2:1317−1327
18. Wilson B et al, Predicting variation in the NVH characteristics of an automatic transmission using a detailed parametric modelling approach. 2007-01-2234
19. Hirabayashi et al (2007) Epicyclic gear transmission error—the importance of controlling tolerances, 2007-01-2241
20. Craig RR, Jr, Bampton MCC (1968) Coupling of substructures for dynamic analysis, AIAA J 6(7)
21. Depenbrock M (1988) Direct self-control (DSC) of inverter-fed induction machine. IEEE Trans Power Electr 3:420–429
22. Hofmann A, Al-Dajani A, Bosing M, De Doncker RW (2013) Direct instantaneous force control: A method to eliminate mode-0-borne noise in switched reluctance machines. In: 2013 IEEE international electric machines drives conference (IEMDC), pp 1009−1016
23. Pellerey P, Lanfranchi V, Friedrich G (2012) Coupled numerical simulation between electromagnetic and structural models. inuence of the supply harmonics for synchronous machine vibrations. IEEE Trans Mag 48(2):983-986
24. Garvey SD (1989) The vibrational behaviour of laminated components in electrical machines. In: Fourth International conference on electrical machines and drives
25. Davis T (2007) Failure mode avoidance, a 1-day course at the University of Bradford
26. Hofmann A, Qi F, De Doncker RW (2014) Developing the concept for an automotive high-speed SRM drive with focus on acoustics. In: 7th IET Int Conf Power Electr Mach Drives (PEMD 2014), 1(5):8–10

Chapter 4
Cylindrical Nearfield Acoustical Holography: Practical Aspects and Possible Improvements

Matteo Kirchner and Eugenius Nijman

Abstract This chapter discusses Nearfield Acoustical Holography (NAH) for the characterization of cylindrical sources. Cylindrical NAH is an experimental airborne characterization technique, and it is suited for any type of cylindrical source. NAH allows to evaluate sound intensity, pressure level and particle velocity. Practical aspects of Nearfield Acoustical Holography such as positioning error, measurement noise, hologram distance and measurement aperture are investigated and discussed with the aid of numerical examples. Moreover, a technique referred to as compressive sampling (CS) is discussed, aiming to reduce the number of sensors required by the classical NAH in the high frequency range.

Keywords NVH · Cylindrical NAH · Holography · Acoustics · Airborne noise · Compressive sampling

4.1 Introduction

This chapter deals with Nearfield Acoustical Holography (NAH) in cylindrical coordinates, which is an experimental method to characterize a cylindrical airborne sound source. Cylindrical NAH deals with a hologram, which is a set of measurements of the sound field in discrete positions on a cylindrical surface around a source. The methodology is generic and can be applied to any cylindrical source.

M. Kirchner (✉) · E. Nijman
Virtual Vehicle Research Center, Area NVH and Friction,
Inffeldgasse 21/a, 8010 Graz, Austria
e-mail: matteo.kirchner@kuleuven.be

E. Nijman
e-mail: eugene.nijman@v2c2.at

M. Kirchner
Department of Mechanical Engineering, KU Leuven, Celestijnenlaan 300,
3001 Heverlee, Belgium

© The Author(s) 2016
A. Fuchs et al. (eds.), *Automotive NVH Technology*,
Automotive Engineering: Simulation and Validation Methods,
DOI 10.1007/978-3-319-24055-8_4

NAH is based on a two-dimensional decomposition of the hologram in spatial Fourier components (wavenumber space, or k-space) [1]. This implies an equally spaced microphone lattice and a dependence on the Nyquist–Shannon sampling theorem, leading to an overwhelming amount of microphones in the high frequency range, limiting the practical application of this method. Moreover, increasing the amount of microphones may modify the sound field around the source (because of the wave-microphone interaction). In a recent publication [2] compressive sampling (CS) has been suggested as a potential solution for this problem. In fact, NAH with compressive sampling has been reported to increase the high frequency limit using less transducers. The formulation of NAH with CS is presented and explained here, together with a few comments about its applicability.

This chapter is structured as follows: first, the classical NAH approach is introduced, followed by the description of the numerical test case that will be adopted to explain a few practical aspects, such as measurement aperture, positioning error and measurement noise. Next, holography is reformulated such that compressive sampling principles can be applied. Finally, a few concerns with regard to the applicability of compressive sampling to NAH will be outlined.

4.2   Cylindrical Nearfield Acoustical Holography

4.2.1   NAH Formulation

Cylindrical Nearfield Acoustical Holography consists in the following four steps:

1. Measurement of the pressure (or alternatively the particle velocity [3, 4]) on a grid of points on the cylindrical hologram surface. The work presented here deals only with pressure transducers, due to their larger availability and lower price.
2. Computation of the spectrum through a spatial transform, which gives a wavenumber space (k-space) description of the original distribution.
3. Multiplication of the spectrum by an inverse propagator for the reconstruction of the pressure- or velocity-distribution on the target surface. If the target surface corresponds to the source surface, the source velocity distribution may be obtained. Only the reconstruction of the surface velocity distribution is addressed here.
4. Back transformation to the real space.

The details about NAH can be found in the book of Earl G. Williams [1]. NAH is prone to instability because of the presence of evanescent waves. In fact, for planar geometries the evanescent waves decay exponentially, while exponential and power law decays are observed for cylindrical and spherical geometries. Those behaviors lead to an ill-posed inverse problem, where they turn into exponential-like amplifications causing the "blowing up" of the noise and inaccuracies contained in the measured data. A regularization procedure needs therefore to be applied in order to stabilize the solution. Among the available regularization methods (generic mathematical tools [5–7] and applications to holography [8]), Tikhonov regularization has been applied to the simulations presented here [9].

NAH can be formulated as a single matrix expression. First, let us consider the forward problem shown in Eq. (4.1) [8], where p is the spatial pressure distribution at a given frequency, that is, the pressures measured in the discrete hologram points, and w is the velocity distribution on the source. H is built from a diagonal matrix G containing the direct propagators of each k-space component, pre-multiplied by an inverse DFT matrix F^{-1}, and post-multiplied by F as shown in Eq. (4.2).

$$p = Hw \qquad (4.1)$$

$$H = F^{-1}GF \qquad (4.2)$$

The inverse problem is shown in Eqs. (4.3) and (4.4), and follows from Eq. (4.1). R_α is the Tikhonov regularized inverse, which depends on the regularization parameter α. The superscript H denotes the Hermitian matrix (conjugate transpose). Note that if $\alpha = 0$ (no regularization), R_α is simply the pseudo-inverse of H.

$$w = R_\alpha \cdot p \qquad (4.3)$$

$$R_\alpha = (H^H H + \alpha \cdot I)^{-1} H^H \qquad (4.4)$$

Among the several methods to seek the value of α that best regularizes R_α, the generalized cross validation (GCV) offers a way to get to a very accurate solution without the need to know the variance of the noise of the system [5, 8, 9].

It is, however, important to emphasize that the GCV scheme is not able to stabilize the problem if the noise components contaminating the measurement data are highly correlated [5].

4.2.2   Description of the Numerical Test Case

A numerical test case has been built in order to investigate practical aspects of cylindrical NAH. It consists of a set of acoustical monopoles which are positioned on a cylindrical grid inside a virtual source surface (Fig. 4.1). The source surface consists of a superellipsoid with smooth, almost cylindrical shape. It has a length $l = 0.3$ m and radius $a = 0.12$ m [9]. The volume velocity distribution of the monopoles was chosen in order to obtain the frequency independent source surface velocity distribution presented in Fig. 4.2. For this purpose, the "Source Simulation Technique" [10] was used. The sound pressure in any position of the radiated field is obtained by straightforward superposition of the monopole fields.

Fig. 4.1 Source layout.
Source surface (*green points*)
and acoustic monopoles
(*black stars*)

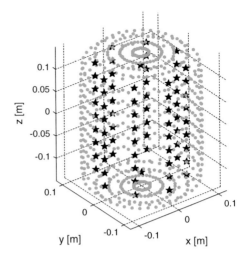

Fig. 4.2 Normal velocity
distribution on the source,
generated by the set of
acoustic monopoles

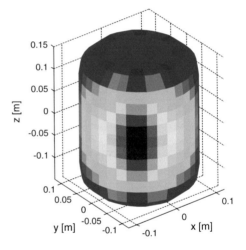

Figure 4.3 shows the hologram, i.e., a set of discrete measurement points on a concentric cylindrical surface enveloping the source. Unless otherwise stated, the length of the hologram is set to 1.5 times the length of the source, and the distance from the hologram to the surface of the source is d = 0.05 m [9]. Moreover, the spatial sampling is chosen in such a way that the unwrapped mesh presents a similar microphone spacing in the two directions (longitudinal and circumferential sampling $\Delta \approx 0.05$ m). Under these spatial sampling conditions, the frequency limit set by the Shannon sampling theorem yields (in air, where the speed of sound is $c_0 \approx 343$ m/s):

Fig. 4.3 Source surface
(*green points*) and hologram
surface (*blue stars*)

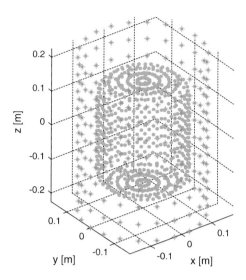

$$f = \frac{c_0}{2\Delta} \approx \frac{343}{2 \cdot 0.05} = 3430\,\text{Hz} \tag{4.5}$$

A finer lattice can be designed if higher frequencies have to be taken into account. However, increasing the amount of microphones may modify the sound field around the source (because of the wave-microphone interaction). Moreover, the data acquisition system may become much more expensive due to the higher number of channels required. These aspects have to be taken into account when designing the microphone antenna.

To show the effects of the regularization, a small error of zero mean and standard deviation $\sigma = 0.25$ dB was added to the simulations. Figure 4.4 shows the effect of the regularization procedure on the surface velocity reconstruction along the z axis at $\theta = 0.29$ rad and along the circumference at $z = 0$ m. The graphs include the velocity distribution created by the acoustic monopoles (dashed green line), that the NAH algorithm seeks to reconstruct. The dotted blue line refers to the backpropagation without any regularization, while the solid blue line illustrates the regularized procedure. The velocity level (VL) is expressed in dB according to the formula of Eq. (4.6), where $v_{ref} = 5 \cdot 10^{-8}$ m/s.

$$VL = 10 \cdot \log_{10}\left(\frac{v^2}{v_{ref}^2}\right) \tag{4.6}$$

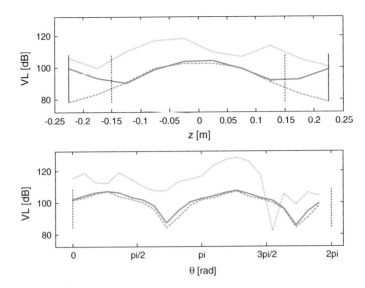

Fig. 4.4 Effect of the regularization procedure on the surface velocity reconstruction along the z axis at $\theta = 0.29$ rad (*top*) and along the circumference at $z = 0$ m (*bottom*). Legend: real velocity distribution (*dashed green*), NAH not regularized (*dotted blue*), NAH regularized with Tikhonov and GCV (*solid blue*). The graphs refer to f = 1010 Hz. The *vertical black lines* in the *top figure* help to visualize the geometry: the *dotted lines* represent the spatial limits of the source, while the *solid lines* are the longitudinal spatial limits of the hologram

The results obtained without regularization are far from being accurate, while the situation completely changes with the regularization. In particular, a good reconstruction of the velocity distribution is observed in the region of the source (between the vertical dotted black lines in Fig. 4.4).

Figure 4.5 shows the spatially averaged square error (ε) of the hologram portion corresponding to the source, for a frequency range from 10 to 3010 Hz, with 100 Hz step. ε is defined in Eq. (4.7), where v_0 represents the actual source velocity. The superscript line indicates spatial average.

$$\varepsilon = 10 \cdot \log_{10}\left(\frac{\overline{v^2}}{\overline{v_0^2}}\right), \tag{4.7}$$

At high frequencies, the wavenumber spectrum components exceed the noise and regularization is not necessary, but at low frequencies the low-pass k-space filter introduced by the regularization proves absolutely necessary and gives excellent results.

Fig. 4.5 NAH spatially averaged square error (ε) of from 10 to 3010 Hz and step 100 Hz. NAH not regularized (*dashed green*), and NAH regularized with Tikhonov and GCV (*solid blue*)

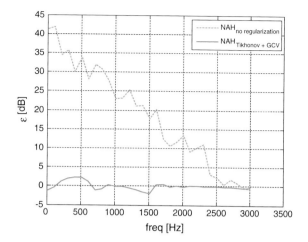

4.3 Practical Aspects of Cylindrical Holography

4.3.1 Hologram Length and Hologram Distance

Considering the case of cylindrical holography, an ideal example that lacks any disturbing element cannot be treated, due to the spectral leakage in the longitudinal direction.

The hologram is closed circumferentially, and an infinitely long periodic signal can thus be recognized along this coordinate. The same concept does not apply to the longitudinal direction, where the DFT creates an infinite series of replicated measurements that extend longitudinally the hologram introducing backpropagations errors.

In order to limit the influence of this phenomenon, the hologram has to be sufficiently long and sufficiently close to the surface of the source, to keep the replicated sources far from the real source. It is consequently important that the microphones which are located nearby the two ends of the hologram measure a much lower pressure than the ones in the central part. On top of this, the hologram distance has to be set as small as possible also in order to detect the exponential decaying evanescent waves. A distance of 5 cm has been chosen here.

Figure 4.6 shows the influence of the hologram length, for a fixed hologram distance and spatial sampling. The curves of the holography error oscillate within 2 dB at low frequency, and become very small above 1700 Hz. As expected, the gray dash dotted line, corresponding to the shortest hologram, leads to the highest error. Since the results of holography performed with a length of 1.5 and 2 times the length of the source are accurate and comparable (solid lines), the first length has been selected as a good choice, limiting the number of microphones.

Fig. 4.6 NAH spatially averaged square error for five different hologram lengths, from 10 to 3010 Hz and step 100 Hz. Hologram distance and spatial sampling are set to 5 cm. l_{ho} is the hologram length, and l is the length of the source

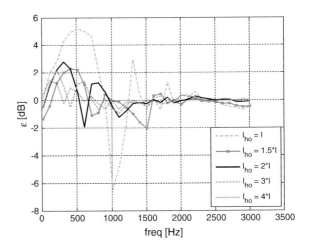

4.3.2 Hologram Positioning Error

In practice, it is possible to create a hologram through either a static microphone array or a single sensor moved by a robot (or alternatively placed manually on a grid), provided that the field is stationary. For both techniques, a certain level of accuracy characterizes the position of the transducers. In the present work, stochastic and systematic positioning errors are distinguished and discussed.

Stochastic errors are due to the accuracy that can be achieved while aligning the acoustic center of the pressure transducers with their theoretical positions on the holographic lattice. This type of error can be effectively corrected by a regularization algorithm.

Furthermore, systematic errors add to the above-mentioned stochastic uncertainty, and consist of centering and alignment errors of the global array as well as its deviation from the circular cross section. Unfortunately, Tikhonov regularization is not able to filter this type of error [5].

First, some simulations aim to show how much stochastic errors influence the reconstruction of an acoustic source. Figure 4.7 shows the spatially averaged square error for a positioning inaccuracy with normal distribution and standard deviations $\sigma_{xyz} = 2, 4, 6, 8$ mm on each Cartesian axis. The error introduced by the highest inaccuracies (6 and 8 mm) starts deviating from the first two cases at approximately 1700 Hz. This trend is justified by the fact that a given positioning error has a stronger influence on a shorter wavelength. Nevertheless, the holography together with the regularization procedure gives excellent results, with a spatially averaged square error within 3 dB.

Further simulations have been carried out to investigate the effects of three types of systematic error, i.e., a global translation of the entire array of 5 mm in all the three Cartesian axes (Fig. 4.8a), an angular misalignment of 0.01 rad of the longitudinal hologram axis (max displacement of 5 mm, Fig. 4.8b), and a stretching of

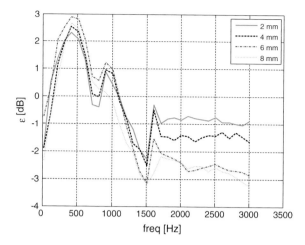

Fig. 4.7 Effect of microphone positioning error for Tikhonov regularized NAH. Spatially averaged square error from 10 to 3010 Hz and step 100 Hz. Standard deviation $\sigma_{xyz} = 2$ mm (*solid black*), 4 mm (*dashed black*), 6 mm (*dash-dotted red*), 8 mm (*dotted gray*)

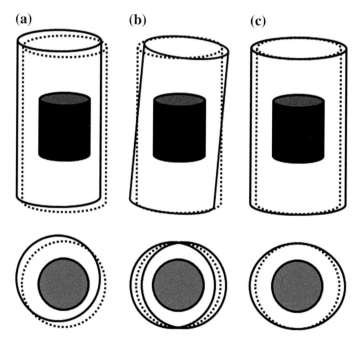

Fig. 4.8 Schematic representation of the systematic errors. Ideal geometry (*dotted*) and geometric error introduced (*solid*)

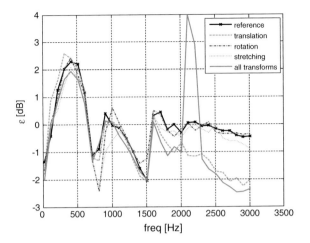

Fig. 4.9 NAH spatially averaged square error regularized with Tikhonov and GCV from 10 to 3010 Hz and step 100 Hz. Effects caused by systematic errors

the circular cylinder section to an oval section (max displacement of 5 mm, Fig. 4.8c). A further simulation has been carried out for the sum of all those systematic errors. Figure 4.9 shows the results. As reference, the case of Fig. 4.5 is also included in the graph.

In general, the errors introduced by this type of inaccuracies are very small or can even be neglected. The magnitude of the error that it introduces in the problem usually does not reach critical values.

4.3.3 Hologram Measurement Noise

Two further causes of noise are discussed here. First, environmental background noise may influence the measurements. Even if it is possible to limit the effect of a disturbing source by isolating it or by calibrating the measurement system, a certain amount of noise will always be present. An acoustic source located outside the holographic lattice may introduce a spatially correlated error which cannot be filtered by the regularization. An example is a wave reflection on a wall. To avoid this disturbing element, free-field conditions are needed. These can be achieved in an anechoic room, where a very high signal-to-noise ratio (SNR) is expected and the influence of an external source such as the ventilation system is extremely low.

The second aspect is related to the electronic noise of the measurement chain. A set of simulations has been performed to investigate this type of noise, showing that the Tikhonov regularization can effectively filter it (it has a spatially uncorrelated shape with an extremely low magnitude).

A further scenario that investigates practical aspects of cylindrical NAH is discussed in Ref. [9]. It refers to the effect of sources which are located on the end shields of the cylinder, and cannot be reached with the backpropagation of a cylindrical contour.

4.4 Possible Improvements: NAH with Compressive Sampling

Promising results have been published with regard to compressive sampling (CS) applied to NAH with the aim to reduce the number of microphones without affecting the quality of the reconstruction [2]. Compressive sampling—sometimes also referred to as compressive sensing, compressed sampling or compressed sensing—is a well known scheme in the field of audio and image processing. This technology allows to slim down data at the source. Compressive sampling wants to directly acquire the minimum amount of data which is needed to fully represent the signal, while in other cases (e.g., in digital photography) a huge amount of data is acquired and immediately compressed in order to fit in a storage drive.

CS is based on a concept referred to as *signal sparsity*, i.e., a signal can be represented (fully or in an approximate way) by just a few components belonging to a certain space. This space is referred to as *dictionary*, and its components are the so-called *basis functions* or *atoms*.

Signals are compressible, i.e., they are well approximated by sparse vectors, when they have a sparse representation in some domain [11–13]. The problem is then to represent a signal in that specific domain and keep only the relevant non-zero elements (the sparser the solution, the better the compression). A simple overview on compressive sampling can be found in [13].

According to the notation that has already been introduced while describing the classical NAH procedure (see Sect. 4.2.1), here the formulation of holography with compressive sampling is given [14].

Formally, the only difference is in the definition of the source velocity distribution w, which is shown in Eq. (4.8). Matrix D represents a dictionary of basis functions (atoms), and α is a complex column vector containing the coefficients of those basis functions.

$$w = D\alpha \qquad (4.8)$$

Dictionaries may be *overcomplete* [15], meaning that the signal description expressed by Eq. (4.8) is not unique, i.e., one basis function can be represented by other basis functions. Mathematically, this corresponds to a non-orthogonal dictionary.

There exist many candidates for building a dictionary for compressing sampling, such as Fourier basis functions, wavelets, chirplets [15]. The crucial point is that the dictionary has to be able to *sparsify* the signal, i.e., only a few atoms should (fully or in an approximate way) describe the signal. Therefore, α should be sparse. The reason why signal sparsity is a crucial aspect in CS will become clear later.

The compressive sampling formulation for NAH is given in Eq. (4.9), and was obtained by substituting Eqs. (4.2) and (4.8) in (4.1).

$$p = F^{-1}GFD\alpha \tag{4.9}$$

It is important to understand the structure of each component of Eq. (4.9). The easiest way is to rewrite it highlighting its structure and writing explicitly the matrix sizes, as shown in Eq. (4.10).

$$\underset{m\times 1}{\{p\}} = \underset{m\times l}{[F_{inv}]} \underset{l\times l}{[G]} \underset{l\times l}{[F]} \underset{l\times n}{[D]} \underset{n\times 1}{\{\alpha\}} \tag{4.10}$$

Here follows the description of each term of Eq. (4.10):

- $\{p\}$ is a column vector of length m containing the measurements, sampled in space. It consists of complex numbers, since each measurement point carries the frequency domain information at a frequency step. The measurement points may or may not be regularly spaced. A randomly spaced hologram helps to capture the high frequency information with fewer measurements in comparison to the number imposed by Nyquist–Shannon sampling theorem [2].
- $\{\alpha\}$ is a column vector of length n with the complex coefficients of the basis functions. It should be as sparse as possible, in order to allow fewer measurement points in comparison with classical NAH. The reason for this is that there is a relationship between sparsity and required number of measurements [12].
- $[D]$ is the dictionary. Its columns are the n basis functions (atoms). Each basis function is evaluated on a set of l regularly spaced points. The choice of l determines the number of spatial Fourier components that will be propagated through $[G]$. The finer the sampling, the higher the number of (evanescent) components taken into account.
- $[F]$ is a square matrix that applies the two-dimensional Fourier transform (2D-DFT) to the basis functions.
- $[G]$ has the same size of $[F]$. It is a diagonal matrix containing the (complex) propagators related to the corresponding Fourier components. The propagators are defined here for the forward problem (exterior problem), i.e., the propagation goes from the source to the (cylindrical) surface on which the hologram is located.
- $[F_{inv}]$ cannot anymore be computed by inverting $[F]$, as it was in Eq. (4.2). This is because it may not be square anymore (typically, $m < l$). Moreover, it may be evaluated on a set of randomly spaced points, accordingly to the m measurement positions.

Equation (4.9) can be rewritten in a more compact version, which highlights the measurements vector p, the (sparse) complex coefficients vector α, and a transform matrix A between them:

$$p = A\alpha, \tag{4.11}$$

where

$$A = F_{inv}GFD. \tag{4.12}$$

Since the general aim of compressive sampling is to reduce the number of measurement points, there will always be more basis functions than measurement points, i.e., $n > m$. Consequently, the system described by Eqs. (4.9) and (4.11) is *underdetermined*, i.e., there are there are fewer equations (measurement points) than unknowns (basis functions).

In order to reconstruct the signal, Eq. (4.11) has to be inverted and computed. The solution has then to be inserted into Eq. (4.8) to obtain the velocity distribution.

Since Eq. (4.11) is an underdetermined system, it has an infinite number of solutions. Among them, the sparsest solution is the best candidate for the success of compressive sampling. The common Moore-Penrose pseudo-inverse (which relies on a L_2-norm optimization, i.e., on the least mean square error) does not address this specification [11].

Given an arbitrary underdetermined system such as Eq. (4.11), the sparsest among all possible solutions is obtained by a minimization of the L_0-norm of the vector α [16]:

$$\min_{\alpha} \|\alpha\|_{\ell_0} \text{ subject to } A\alpha = p. \tag{4.13}$$

Unfortunately, there are no efficient algorithms to solve such problem, due to the non-convexity of the L_0-norm optimization [16, 17]. Fortunately, those limitations can be overcome by "relaxing" the L_0-norm up to an L_1-norm problem [18, 19], for the solution of which some methods are available.

The so-called "basis pursuit" (BP) has recently been strongly developed, and efficient algorithms are now available [15]. Basis pursuit is capable to find the sparsest solution among a dictionary composed by non-orthogonal basis functions [15, 20]. Moreover, BP is based on global optimization, offers good sparsity and stable superresolution, and can be used with noisy data [15].

Under a condition referred to as Restricted Isometry Property (RIP) [12, 16, 21], both the L_0-norm and L_1-norm problems are proven to give the same and unique result.

The relaxation from L_0-norm to L_1-norm of Eq. (4.13) leads to the new problem described by Eq. (4.14), the solution of which will be sought through a basis pursuit algorithm [15].

$$\min_{\alpha} \|\alpha\|_{\ell_1} \text{ subject to } A\alpha = p. \tag{4.14}$$

A detailed description of all aspects that have to be dealt with when employing compressive sampling for Nearfield Acoustical Holography would go beyond the purpose of this chapter. The interested reader can refer to [14]. Here, just a few elements of concern are listed.

First, it should be mentioned that it is not easy to have the RIP satisfied while performing NAH. This fact is linked to the effect of the NAH propagators (matrix $[G]$). The RIP characterizes matrices which are nearly orthonormal, at least when operating on sparse vectors. This condition cannot hold when propagating and evanescent waves are present at the same time [14].

Another aspect regards the sparsity of the source, i.e., the number of nonzero elements of a sparse signal. If this number can be estimated accurately, then it is possible to design an acquisition system with a minimal amount of sensors. This can happen only if a certain knowledge of the source is known a priori, such that a dictionary that promotes sparsity can be chosen.

In other words, compressive sampling helps reducing the number of sensors required in the high frequency range only for small hologram distances and if the sparsity of the source can be predicted accurately [14].

4.5 Conclusions

This chapter investigated a few practical aspects of cylindrical Nearfield Acoustical Holography (NAH) and discussed compressive sampling (CS) applied to NAH in order to reduce the number of microphones needed in the high frequency range.

A few numerical examples involving cylindrical NAH have been shown, and the formulation of NAH with compressive sampling has been presented and discussed.

The following conclusions could be drawn:

- Tikhonov regularization combined with a Generalized Cross Validation parameter selection effectively improves the results of NAH.
- The proposed antenna layout allows for microphone positioning errors up to 6 mm without introducing substantial errors ($\varepsilon \leq \pm 3$ dB) provided that the over mentioned regularization scheme is employed.
- The distance from the hologram to the source shall be as small as possible. Typically, the hologram length shall exceed the source longitudinal dimension by a factor of at least 1.5.
- Tikhonov regularization is not able to filter systematic noise. Fortunately, the impact of the systematic errors on the velocity reconstruction turns out to be marginal. A structure whose systematic errors do not affect the position of a microphone more than 5 mm (evaluated as maximum distance to the ideal position of a hologram in case of rigid translation, rotation or stretch) causes no detectable changes on the averaged square error. External sources shall be avoided by choosing anechoic test environments with low background noise.
- Compressive sampling is based on a concept referred to as signal sparsity and relies on a matrix condition called Restricted Isometry Property (RIP). If the RIP holds, the number of transduces needed to correctly acquire the signal is proportional to the sparsity. Unfortunately, the propagators of the wavenumber

spectrum (k-space) deteriorate the RIP, especially when the different amplification of propagating and evanescent waves becomes substantial.

- The application of compressive sampling to Nearfield Acoustical Holography is promising only if the sparsity of the acoustic field is known. In other words, the goal of having less microphones can be achieved only if the sparsity of the signal is high and can be estimated. For the characterization of an arbitrary unknown acoustic source, this seems to be the most critical step to be overcome.

Acknowledgement The research work of Matteo Kirchner has been funded by the European Commission within the FP7 EID Marie Curie project "eLiQuiD" (GA 316422). The authors acknowledge the financial support of the "COMET K2—Competence Centres for Excellent Technologies Programme" of the Austrian Federal Ministry for Transport, Innovation and Technology (BMVIT), the Austrian Federal Ministry of Science, Research and Economy (BMWFW), the Austrian Research Promotion Agency (FFG), the Province of Styria and the Styrian Business Promotion Agency (SFG). The IWT Flanders and KU Leuven research fund are also gratefully acknowledged for their support. Finally, the authors gratefully acknowledge the support of COST action TU1105.

References

1. Williams EG (1999) Fourier acoustics: sound radiation and nearfield acoustical holography. Academic Press, London
2. Chardon G, Daudet L, Peillot A, Ollivier F, Bertin N, Gribonval R (2012) Near-field acoustic holography using sparse regularization and compressive sampling principles. J Acoust Soc Am (JASA) 132(3):1521–1534
3. Jacobsen F, Liu Y (2005) Near field acoustic holography with particle velocity transducers. J Acoust Soc Am 118(5):3139–3144
4. Ollivier F, Le Moyne S, Picard C (2007) Experimental comparison of pu probes and microphone arrays used in impulse acoustic holography. In: Proceedings of 14th international congress on sound & vibration (ICSV14)
5. Hansen PC (1998) Rank-deficient and discrete ill-posed problems: numerical aspects of linear inversion. Society for Industrial and Applied Mathematics (SIAM)
6. Nelson PA, Yoon SH (2000) Estimation of acoustic source strength by inverse methods: part I. Conditioning of the inverse problem. J Sound Vib 233(4):639–664
7. Yoon SH, Nelson PA (2000) Estimation of acoustic source strength by inverse methods: Part II. Experimental investigation of methods for choosing regularization parameters. J Sound Vib 233(4):665–701
8. Williams EG (2001) Regularization methods for near-field acoustical holography. J Acoust Soc Am 110(4):1976–1988
9. Kirchner M, Nijman E (2014) Nearfield acoustical holography for the characterization of cylindrical sources: practical aspects. In: 8th international styrian noise, vibration & harshness congress (ISNVH 2014), Graz, Austria
10. Ochmann M (1995) The source simulation technique for acoustic radiation problems. Acta Acustica United Acustica 81(6):512–527
11. Baraniuk R (2007) Compressive sensing [lecture notes]. IEEE Signal Process Mag 24(4): 118–121
12. Candes E, Wakin M (2008) An introduction to compressive sampling. IEEE Signal Process Mag 25(2):21–30

13. Hayes B (2009) The best bits. Am Sci 97(4):276–280
14. Kirchner M, Nijman E (2014) Cylindrical nearfield acoustical holography using compressive sampling: feasibility and numerical examples. In: Proceeding of ISMA2014 including USD2014. Leuven, Belgium, 15–17 Sept 2014, pp 1531–1546
15. Chen SS, Donoho DL, Saunders MA (2001) Atomic decomposition by basis pursuit. SIAM Rev 43(1):129–159
16. Candes E, Tao T (2005) Decoding by linear programming. IEEE Trans Inf Theory 51 (12):4203–4215
17. Candes E, Romberg J, Tao T (2006) Robust uncertainty principles: exact signal reconstruction from highly incomplete frequency information. IEEE Trans Inf Theory 52(2):489–509
18. Donoho D (2006) Compressed sensing. IEEE Trans Inf Theory 52(4):1289–1306
19. Donoho D, Tanner J (2006) Thresholds for the recovery of sparse solutions via l1 minimization. In: 40th annual conference on information sciences and systems, pp 202–206
20. Chen S, Donoho D (1994) Basis pursuit. In: 1994 conference record of the twenty-eighth asilomar conference on signals, systems and computers, vol 1, pp 41–44
21. Candes E (2008) The restricted isometry property and its implications for compressed sensing. CR Math 346(910):589–592

Chapter 5
Vibro-Acoustic Analysis of Geared Systems—Predicting and Controlling the Whining Noise

Alexandre Carbonelli, Emmanuel Rigaud and Joël Perret-Liaudet

Abstract The main source of excitation in gearboxes is generated by the meshing process. It is usually assumed that static transmission error (STE) and gear mesh stiffness fluctuations are responsible of noise radiated by the gearbox. They generate dynamic mesh forces which are transmitted to the housing through wheel bodies, shafts and bearings. Housing vibratory state is directly related to the noise radiated from the gearbox (whining noise). This work presents an efficient method to reduce the whining noise The two main strategies are to reduce the excitation source and to play on the solid-borne transfer of the generated vibration. STE results from both tooth deflection (depending of the teeth compliance) and tooth micro-geometries (voluntary profile modifications and manufacturing errors). Teeth compliance matrices are computed from a previous finite elements modeling of each toothed wheel. Then, the static equilibrium of the gear pair is computed for a set of successive positions of the driving wheel, in order to estimate static transmission error fluctuations. Finally, gear mesh stiffness fluctuations is deduced from STE obtained for different applied loads. The micro-geometry is a lever to diminish the excitation. Thus, a robust optimization of the tooth profile modifications is presented in order to reduce the STE fluctuations. The dynamic response is obtained by solving the parametric equations of motion in the frequency domain using a spectral iterative scheme, which reduces considerably the computation time. Indeed, the proposed method is efficient enough to allow a dispersion analysis or parametric studies. The inputs are the excitation sources previously computed and the modal basis of the whole gearbox, obtained by a finite element method and including gears, shafts, bearings and housing. All the different parts of this global approach have been validated with comparison to experimental data, and lead to a satisfactory correlation.

A. Carbonelli (✉)
Vibratec, Ecully Cedex, France
e-mail: alexandre.carbonelli@vibratec.fr

E. Rigaud · J. Perret-Liaudet
Ecole Centrale de Lyon - LTDS, Ecully Cedex, France
e-mail: emmanuel.rigaud@ec-lyon.fr

J. Perret-Liaudet
e-mail: joel.perret-liaudet@ec-lyon.fr

© The Author(s) 2016
A. Fuchs et al. (eds.), *Automotive NVH Technology,*
Automotive Engineering: Simulation and Validation Methods,
DOI 10.1007/978-3-319-24055-8_5

63

Keywords Vibro-acoustic · Gear mesh dynamics · Gear optimization · Whining noise

5.1 Introduction

Geared systems are the seat of vibrations induced by the meshing process. For this reason, a gearbox is an important source of noise and vibration in automotive industry. The gearbox internal sources of excitation are various. The main source corresponds to fluctuation of the static transmission error (STE) of the gear which transmits the drive torque [1, 2]. STE corresponds to the difference between the actual position of the driven gear and its theoretical one.

$$STE(\theta_1) = R_{b2}.\theta_2(\theta_1) - R_{b1}\theta_1 \qquad (5.1)$$

where R_{bj}, is the base radius of gear j.

It is mainly due to voluntary (corrections) and involuntary (defects) geometrical deviations of the teeth at a micrometric scale and to elastic deformation of loaded teeth, wheel bodies and crankshafts. STE fluctuations also generates mesh stiffness fluctuations. Under operating conditions, the parametric excitations induce dynamic loads at the gear meshes, which are transmitted to the gearbox receiving structure via the wheel bodies, crankshafts and bearings, as presented in Fig. 5.1. The vibratory state of the crankcase is the main source of the radiated noise [3].

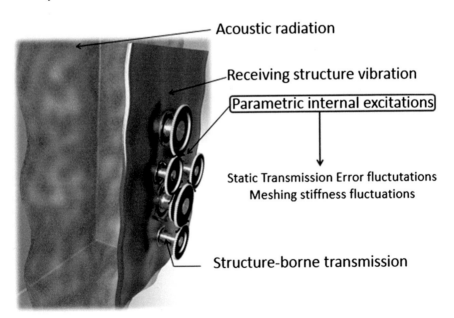

Fig. 5.1 Whining noise generation process

STE fluctuations need to be minimized by introducing voluntary tooth micro-geometrical modifications in order to reduce the radiated noise. For this study, the selected optimization parameters for each gear pair are:

- the tip relief values X of pinion and driven gear i.e. the amount of material remove on the teeth tip,
- the starting tip relief diameters ϕ of pinion and driven gear,
- the added up crowning centered on the active tooth width $C_{\beta,i/j}$.

The optimization of tooth modifications in simple mesh gear system for a given torque has been studied by many authors [4–6] but the approach for multi-mesh gear systems optimization is still unusual [7]. The first part of this paper presents a complete optimization process for a truck timing cascade of gears displayed in Fig. 5.2.

In this study, the first cascade is designed with three helical gears and has 8 optimization parameters (2 by gear, and 1 by mesh). The second cascade is designed with two gears and has therefore 5 optimization parameters. Moreover the modifications made on teeth profile have to be satisfying for a wide torque range. That requires an efficient method as the number of possible solutions is extremely large, due to the combinatory explosion phenomenon. The Particle Swarm Optimization (PSO) [8] has been chosen because it is particularly efficient as it is an order 0 meta-heuristic, i.e. it not necessary to evaluate first derivatives of the function.

Furthermore, robustness of the obtained solutions has to be studied. Indeed, dispersion of manufacturing errors generates a strong variability of the dynamic behavior and noise radiated from geared systems (sometimes up to 10 dB [9, 10]).

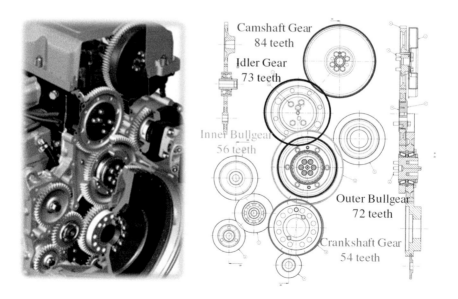

Fig. 5.2 Studied gears for STE computation and optimization

| Primary and Secondary shafts mesh | Secondary and Differential shafts mesh |

Fig. 5.3 Studied gearbox for the dynamic response computation

A statistical study of solutions permits to have a good overview of how the solution can be deteriorated when the manufacturing errors (dispersion over the optimization parameter values) and assembling errors (lead summed up and involute alignment deviations, respectively $f_{H\beta}$ and $f_{g\alpha}$) are considered.

The dynamic response computation procedure is applied to an automotive gearbox displayed in Fig. 5.3.

This computational scheme requires a finite element model of the complete gearbox in order to obtain its modal basis. The contact between the gears is modelled with a stiffness matrix linking the degrees of freedom of each pair of meshing gears. To achieve that, the mean value of the mesh stiffness is taken, leading to mean modal characteristics. The parametric mesh stiffness $k(t)$ isdirectly related to the applied torque T and the static transmission error $STE(t)$ with:

$$k(t) = \frac{1}{R_b} \frac{\partial T}{\partial STE(t)} \tag{5.2}$$

The scheme uses then a powerful resolution algorithm in frequency domain to solve the dynamic equations with an iterative procedure [11, 12]. The original spectral iterative method has been extended in order to take into account several parametric excitations [13]. In that case, there is a coupling between the excitations due to the stiffnesses fluctuations. The corresponding set of equations for m meshes is then.

$$M\ddot{x} + C\dot{x} + Kx + \sum_{j=1}^{m} k_j(t) R_j R_j^T x = \sum_{j=1}^{m} k_j R_j STE_j(t) \tag{5.3}$$

M, C, and K   are respectively the global mass, damping and stiffness matrices of the system.

x   the vector of the generalized coordinates of the system, (˙) stands for the time derivative.

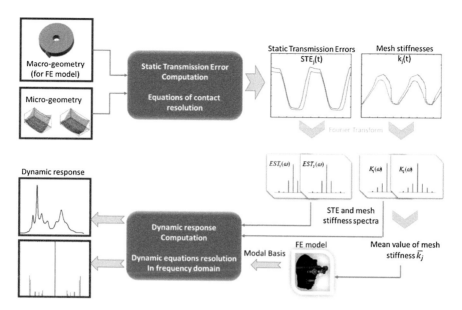

Fig. 5.4 Overview of the computational scheme

R_j is a vector of macro-geometric coupling of the degrees of freedom of two meshing gears.

The final outputs are the housing vibration as a function of the frequency. The operating speeds corresponding to resonance peaks and vibration amplitudes of the housing characterize the whining noise severity. The process can be repeated for several applied torques and can be used to optimize the other gearbox components (for instance the geometry of the housing to minimize its vibration, stiffness of gear bearings...) or to test many different STE from different teeth geometry.

All the computational scheme steps are summarized in Fig. 5.4.

5.2 Static Transmission Error Computation and Robust Optimization of Tooth Profile Modifications

An optimization problem requires a correctly defined fitness function and an appropriate algorithm to be solved. This part describes the choices made to handle this problem considering all the difficulties. Moreover, the robustness study approach is also detailed.

5.2.1 Static Transmission Error Computation

The method for STE calculation retained is classical [14, 15]. Equations describing contact between gears arc solved for each meshing position, taking account of the elasto-static deformations and initial gaps between teeth surfaces.

5.2.2 Optimization Fitness Function Establishment

The criterion retained to estimate one STE fluctuations is the peak-to-peak amplitude (STEpp). Considering that the modifications made have to reduce the STEpp for a given $[T_{min} - T_{max}]$ torques range, the fitness function f is defined as the integral of STEpp over this torques range approximated by a 3-points Gaussian quadrature:

$$f = \int_{T_{min}}^{T_{max}} p(T)STE_{pp}(T)dT \approx \frac{1}{2}\sum_{i=1}^{3} a_i STE_{pp}(T_i) \qquad (5.4)$$

The torque distribution function p(T) is assumed uniform: $p(T) = \frac{1}{T_{max}-T_{min}}$.

a_i are the Gaussian weighting coefficients with the following values: $a_1 = a_3 = 0.5556$ and $a_2 = 0.8888$.

$$\begin{cases} T_1 = \frac{T_{max}-T_{min}}{2} \left(1 - \sqrt{\frac{3}{5}}\right) \\ T_2 = \frac{T_{max}-T_{min}}{2} \\ T_3 = \frac{T_{max}-T_{min}}{2} \left(1 + \sqrt{\frac{3}{5}}\right) \end{cases}$$

C_i are the Gaussian points located in the following way: For the first three-gears-cascade, the multi-objectives aspect is simply handled by considering:

$$f_{84/73/56} = \frac{1}{2}\sum_{i=1}^{3} a_i\left(STE_{pp,83/73}(T_i) + STE_{pp,73/56}(T_i)\right) \qquad (5.5)$$

The second fitness function associated with the other meshing gears is then directly:

$$f_{54/72} = \frac{1}{2}\sum_{i=1}^{3} a_i\left(STE_{pp,54/72}(T_i)\right) \qquad (5.6)$$

5.2.3 Particle Swarm Optimization

The method is based on a stigmergic behavior of a population, being in constant communication and exchanging information about their location in a given space to determine the best location according to what is being searched. In this case, some informant particles are considered, which are located in an initial and random position in a hyper-space built according to the different optimization parameters. The best location researched is thus the combination of parameters which ensures the minimum value of the fitness function defined earlier. At each step and for each particle i, a new speed $V_i(t)$ and so a new position $p_i(t)$ is reevaluated considering:

- the current particle velocity $V_i(t-1)$
- its current position $p_i(t-1)$,
- its best position $p_{i(ind.)}$
- the best position of neighbors $p_{glob.}$.

The algorithm working can be summarized to the system of equations (5.7).

$$
\begin{aligned}
V_i(t) &= \varphi_0 V_i(t-1) + \varphi_1 A_1 \left[p_{i(ind.)} - p_i(t-1) \right] + \varphi_2 A_2 \left[p_{glob.} - p_i(t-1) \right] \\
p(t) &= p_i(t-1) + V_i(t)
\end{aligned}
\tag{5.7}
$$

A_1 and A_2 are random vectors of numbers between 0 and 1 and the coefficients φ_i are taken following Clerc and Trelea [16, 17] works:

$$
\varphi_0 = 0.729 \text{ and } \varphi_1 = \varphi_2 = 1.494.
$$

5.2.4 Robustness Statistical Study

Let's say that a solution S_0 is determined by the PSO. The robustness study is done using a Monte–Carlo simulation, i.e. 10,000 others solutions are computed, chosen randomly in an hyperspace centered on the optimized solution parameters values, limited by the tolerances interval of each parameter and considering possible lead and involute alignment deviations. These 10,000 results allow the establishment of the density probability function of each selected optimized solution. They also allow us to compute statistical variables such as mean value and standard deviation. For statistically independent variables, the theoretical convergence on these values is proportional to $n^{-1/2}$, where n is the number of solutions computed. The industrial request is to consider that the distribution functions of all parameters and errors should be taken uniform. The convergence has been tested and confirmed this convergence law. Therefore the number of samples for a Monte–Carlo simulation has been set to 10,000 ensuring an error less than 1 %.

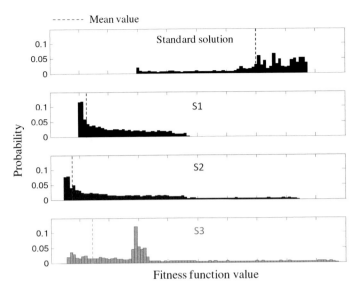

Fig. 5.5 Probability density functions for the standard solution and three selected optimized solutions for the three-gears-cascade

Figure 5.5 shows an example of probability density functions for different possible solutions. It illustrates how the optimized solution is selected. The solution S2 has a smaller mean value, but it is associated to a larger dispersion. The solution S1 appears to be the best compromise between the mean value and the deterioration capability of the solution.

5.2.5 Results—Expected Static Transmission Errors and Actual Ones

After considering the STE_{pp} and its robustness, optimized solutions have been retained for the first three-gears-cascade and for the second two-gears mesh. The evolution of the STE_{pp} is calculated as a function of the applied torque for the standard and the optimized sets of gears. Some measurements have been done to determine the actual teeth topologies, allowing the confrontation of the recommendations made and the tooth modifications obtained. This permits to underline the robustness study pertinence, especially in this study where the (confidential) tolerance intervals are of the same order of magnitude as the tooth modifications themselves.

5.2.6   Theoretical Versus Actual Tooth Topologies

The topology measurements permit to compare the recommended tooth modifications to the actual performed ones, for both standard (corrected but not optimized) and optimized tooth modifications. The whole data are presented in Fig. 5.6 for the crankshaft gear. Figure 5.7 displays only the discrepancies between theoretical and actual tooth surfaces for the all the studied gear. The results plotted are the mean value of all teeth topologies for a given gear. The analysis of these topologies leads to the following observations:

- There are relatively important discrepancies between theoretical and actual teeth topologies.
- The largest discrepancies correspond to the idler and bull inner gears for optimized solutions.

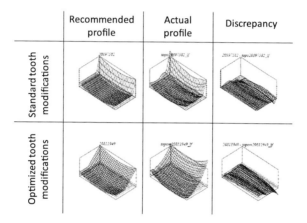

Fig. 5.6 Comparisons of theoretical and actual teeth mean topologies of the crankshaft gear for standard and optimized solutions

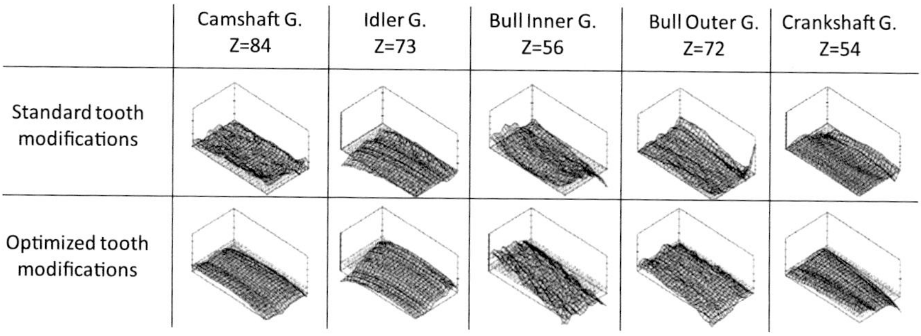

Fig. 5.7 Comparisons of theoretical and actual teeth mean topologies of the studied gears for standard and optimized solutions

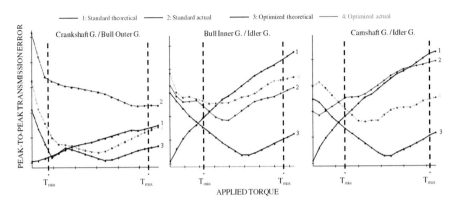

Fig. 5.8 Comparisons of theoretical and actual STE_{pp} as a function of the applied torque for the three meshes

- There are manufacturing errors which cannot easily be translated in terms of tip relieves and crowning.

Figure 5.8 displays the peak-to-peak transmission errors for the three considered meshes. Two results are particularly relevant. The deterioration of theoretical configuration is coherent with the discrepancies presented in Fig. 5.7: the Idler Gear/Bull Inner Gear mesh for the optimized tooth modifications is the mesh with the largest discrepancies. Indeed the corresponding STE_{pp} is worse than the standard actual STE_{pp}. For the other meshes, the curve indexed 2 and 4 shall be compared. The robust optimization done is thus efficient as the STE_{pp} is lower for optimized solutions for the whole torque range.

5.2.7 Acoustic Benefits of the Optimization

Both standard and optimized gears sets have been mounted on a thermal engine and the corresponding radiated noise has been measured. Results are plotted in Fig. 5.8. The benefit is less than expected but the operating torque in a little bit higher than T_{max} and the complete timing system had to be considered (e.g. the oil pump pinion is necessary for the engine oil supply). Nevertheless, the measurements show at least 1 dB of total power reduction, which is satisfying given that the levels (confidential) are initially not high, that only 5 among 10 pinions have been optimized and that all the other acoustic sources are present during the measurements. It is worth underlying that only the pinions have changed between the two different tests who gave the results of Fig. 5.9 and that on some partial sound power measurements the gain was up to 4 dB. The satisfactory benefits of the optimization have led to making those optimized corrections as the new standard ones for Renault Trucks.

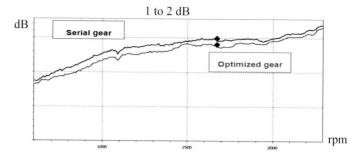

Fig. 5.9 Total power in function of the engine rotation speed

5.3 Dynamic Response Computational Scheme Validation

The computation scheme has been validated step by step by comparison with extensive and complex measurements on a modified but representative automotive gearbox as presented in Fig. 5.10.

Four quantities have been measured: the static transmission error fluctuation, the dynamic transmission error, housing vibration and whining noise. Accelerometers, microphones and optical encoders are used for that purpose. In this paper, the results are mainly focused on the housing vibration.

The measurements were performed at RENAULT's workshop in Lardy in France, on the BACY acyclism test bench. An electrical motor drives the gearbox,

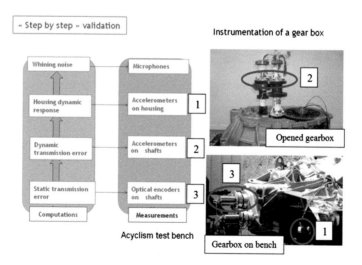

Fig. 5.10 Step by step validation of the computation scheme. Intermediate quantities such as transmission errors, as well as housing dynamic vibration and noise response are compared

while a braking torque simulates the reaction of the wheels. The rotation speed and the torque are also measured.

The tuning of the assembled gearbox has not been done ideally because the experimental modal analysis was not possible on the BACY test bench when a static torque is applied. Preloads effect on mesh and bearings stiffness's are thus not taken into account and even if they may not be negligible [18].

Results from a former experimental modal analysis performed by Vibratec have instead been used. Measurements have been done with a static torque applied but the clamping conditions of the gearbox are different from the ones in the test bench BACY on which the vibration measurements have been done.

As some parts of the gearbox can show non negligible discrepancies over some frequency range, the final assembly is not completely able to describe accurately the dynamic behavior of the measured gearbox. But the results obtained are precise enough to validate the computational scheme.

Figure 5.11 displays the comparison between measured and computed transmission errors. The mean value cannot be obtained by measurements, but the fluctuations, which are the most important data, can be compared. The peak-to-peak amplitude is correctly estimated. The measurement of the transmission error is particularly complicated. The dispersion due to manufacturing errors and assembling errors can be large. Moreover, the micro-geometry should be accurately measured tooth by tooth in order to have real tooth topologies. The agreement between measurements and computations is really satisfying.

The Fig. 5.12 shows a comparison of the housing vibration (dynamic acceleration) as function of the operating speed. The comparison is based on predominance of orders and modes, in terms of frequency and amplitude. The dominant orders and the frequency ranges exhibiting a dynamic amplification correctly determined.

An order tracking has also been done in order to compare properly the vibration measurements with the computations.

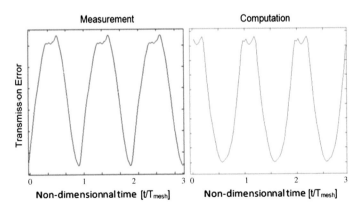

Fig. 5.11 Measured (*left*) and computed (*right*) static error transmission. Peak-to-peak values are compared for the validation (both scales are the same)

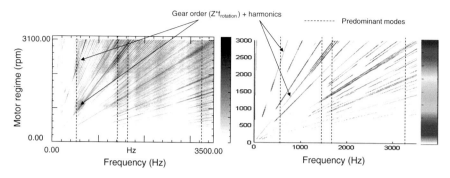

Fig. 5.12 Comparison of measured (*left*) and simulated (*right*) housing vibration as a function of the operating speed. The analysis highlights the principal orders and the predominant frequencies

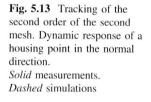

Fig. 5.13 Tracking of the second order of the second mesh. Dynamic response of a housing point in the normal direction.
Solid measurements.
Dashed simulations

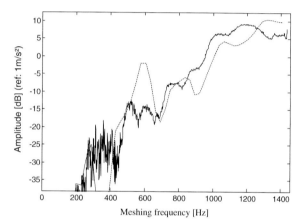

The first and second orders of the two meshes have been considered (the first mesh corresponds to $Z1/Z2 = 35/39$. The second mesh corresponds to $Z3/Z4 = 16/69$ as specified in Fig. 5.3). The acceleration of one housing point for the second order of the second mesh is displayed in Fig. 5.13.

The dynamic model has been tuned in different operating conditions explaining some non-negligible frequency shifts and modal response differences. However the agreement between the measurements and the computations remains satisfying for a predicting tool.

On the contrary to the measurements, the simulation can take into account the variability of the results. Extracted from teeth metrology, a dispersion study has been performed to determine the envelope of the dynamic response. As the teeth micro-geometric dispersion doesn't follow a Gaussian law, the gear defects repartition over the tolerance range has been considered uniform, which is a rather a

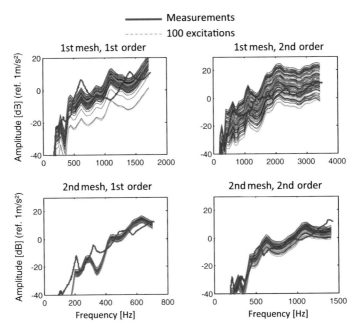

Fig. 5.14 Quadratic mean value over 10 housing points order by order and mesh by mesh. Comparison between measurements and 100 dynamic responses generated randomly according the gear defects dispersion

pessimistic situation. It is well known that the dispersion due to manufacturing and assembling errors can lead strong variability of the dynamic behavior and noise radiated for geared systems. A hundred of excitations have thus been computed and the corresponding responses are plotted in Fig. 5.14 for both meshes and for the two first orders. Discrepancies can be observed due to a bad modal behavior representation at some frequencies. Nevertheless, the order of magnitude of the response is in a good agreement with the measurements. The dispersion for the first mesh is much higher than for the second mesh, and second orders are more sensitive to the dispersion than the first orders.

One of the main industrial concerns is to build a source noise hierarchy to determinate for instance which housing point has the highest vibration level, and which order is dominating.

Figure 5.15 displays the RMS value of the acceleration of some strategically chosen points on the housing, for both simulation and measurements. Considering these results, the computations and the measurements indicate the point n°2 is the less vibrating, and should therefore be used as an attachment point for to the rest of the structure. Once again, the frequency shifts due to a model tuning in different operating conditions introduce some level discrepancies, but the hierarchy between the different points remains suitable as a predictive analysis tool.

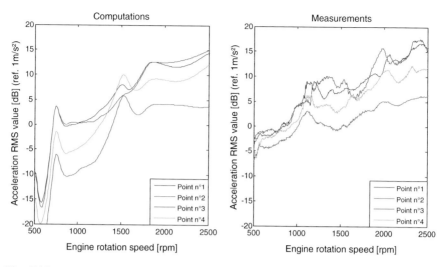

Fig. 5.15 RMS value of the acceleration of some housing points. *Left* simulations. *Right* measurements

5.4 Conclusion

This paper presents a complete approach for whining noise prediction and mini-mization. First, an accurate procedure for computing static transmission error is used. That has been validated over more than 20 cases. The procedure takes into account multi-mesh gear systems and deals with a torque ranges efficiently.

The robustness study is the major contribution, as it is crucial to consider the manufacturing tolerances to find the optimized solution which ensure the best gain. The measurements made on the optimized gears are very encouraging because they lead to a total power diminution of at least 1 dB, even if only 5 among 10 pinions have been optimized. It's worth remembering that all the others acoustic sources were present (thermal engine, accessories...).

The benefit exhibited in the end is satisfying enough to decide to make the optimized corrections the new standard ones for the next Renault Trucks' timing gear cascade.

The dynamic response of a geared system is computed using a spectral iterative scheme, which leads to computation times low enough to permit parametric or dispersion studies. A complete method is proposed to predict whining noise severity, accounting for the scattering of the manufacturing data. The scheme is globally validated and can be used to optimize the current studied gearbox. The computations provide a good estimation of the vibratory response amplitude. They allow identifying the key parameters in order to minimize the whining noise for given functioning configurations, the predominant orders, the critical operating speeds and the accurate hierarchy of these important data. A particular effort should be done to ensure a good numerical model tuning for the prediction remains satisfactory.

Acknowledgement This work was supported by the French National Research Agency through the research project MABCA (ANR 08-VTT_07-02). The partners involved were VIBRATEC, LTDS-Ecole Centrale de Lyon, RENAULT and RENAULT TRUCKS. The authors want to thank especially J. Vialonga from Renault technical center of Lardy (France) and D. Barday from Renault Trucks for their scientific and technical supports and for the shared data.

The French National Research Agency has also supported through the joint laboratory LADAGE (ANR-14-Lab6-003) issued from the collaboration between LTDS-Ecole Centrale de Lyon and VIBRATEC.

References

1. Harris LS (1958) Dynamic loads on the teeth of spur gears. Proc Inst Mech Eng 172:87–112
2. Remond D, Velex P, Sabot J et al (1993) Comportement dynamique et acoustique des transmissions par engrenages. Synthèse bibliographique
3. Welbourn DB (1979) Fundamental knowledge of gear noise—a survey. Proc Conf Noise Vib Eng Trans C177/79:9–29
4. Tavavoli MS et al (1986) Optimum profile modifications for the minimization of static transmission errors of spur gears. J Mech Trans Autom Des 108:86–95
5. Beghini M et al (2004) A method to define profile modification of spur gear and minimize the transmission error. In: Proceedings AGMA fall meeting
6. Bonori G, Barbieri M, Pellicano F (2008) Optimum profile modifications of spur gears by means of genetic algorithms. J Sound Vib 313(3–5):603–616
7. Carbonelli A et al (2011) Particle swarm optimization as an efficient computational method in order to minimize vibrations of multi-mesh gears transmission. In: 2011 advances in acoustics and vibration
8. Eberhart RC, Kennedy J (1995) A new optimizer using particle swarm theory. In: Proceedings of the sixth international symposium on micro machine and human science. IEEE Service Center. Piscataway, pp 39–43
9. Nonaka T, Kubo A, Kato S, Ohmori T (1992) Silent gear design for mass produced gears with scratters in tooth accuracy. In: ASME proceedings of the international power transmission and gearing conference, Scottdale, USA, vol 2, pp 589–595
10. Driot N, Rigaud E, Sabot J, Perret-Laudet J (2001) Allocation of gear tolerances to minimize gearbox noise variability. Acustica United Acta Acustica 87:67–76
11. Perret-Laudet J (1992) Etude des Mécanismes de Transfert entre l'Erreur de Transmission et la Réponse Dynamique des Boîtes de Vitesses Automobiles. Thèse de doctorat de l'Ecole Centrale de Lyon N°9207″
12. Perret-Laudet J (1996) An original method for computing the response of a parametrically excited forced system. J Sound Vib 196:165–177
13. Carbonelli A (2008) Caractérisation vibro-acoustique d'un cascade de distribution poids lourd. Thèse de doctorat de l'Ecole Centrale de Lyon N°2012-34″
14. Rigaud E, Barday D (1998) Modeling and analysis of static transmission error of gears: effect of wheel body deformation and interactions between adjacent loaded teeth. Mécanique Industrielle et Matériaux. 51(2):58–60
15. Rigaud E, Barday D (1999) Modelling and analysis of static transmission error. Effect of wheel body deformation and interactions between adjacent loaded teeth. In: 4th world congress on gearing and power transmission, Paris, vol 3, pp 1961–1972

16. Trelea IC (2003) The particle swarm optimization algorithm: convergence analysis and parameter selection. Inf Process Lett 85(6):317–325
17. Clerc M (1999) The swarm and the queen: towards a deterministic and adaptive particle swarm optimization. In: Proceedings of ICEC, Washington, pp 1951–1957
18. Åkerblom M, Sellgren U (2008), Gearbox noise and vibration—influence of bearing preload, MWL, Department of Vehicle Engineering, KTH, SE–100 44 Stockholm urn University, Auburn, Alabama 36849, USA

Chapter 6
Possibilities and Constraints for Lightweight in Exhaust Systems

Dennis Bönnen, Djahanchah Bamdad-Soufi, Hannes Steinkilberg and Kwin Abram

Abstract In recent years the automotive industry has been using an increasing number of high powered engines with fewer cylinders, with the goal to reduce weight and fuel consumption and hence to achieve lower CO2 emissions. Following, an overview about the currently existing methods and products within the exhaust development is given which follow automotive lightweight trend. Continuous innovations in new materials, structural design and manufacturing process as well as mastering the integration of the components and modules within the system with a thorough understanding and optimization of the system behavior is enabling the reduction of weight in exhaust system. Another possibility to reduce the weight is the use of additional components such as valves. In the following, a discussion about the different types of valves is presented. These valves can be implemented within the exhaust system in order to bring a constraint in the system and consequently additional acoustic damping. Due to engine downsizing, many premium vehicles lost their class-representing sound signature. An active system can be used in order to enhance the sound according to the customer demands. In addition to that, an active system can help reducing muffler volume.

Keywords Exhaust · Silencer · Valve · Active noise · Sound radiation · Lightweight

D. Bönnen (✉) · D. Bamdad-Soufi · H. Steinkilberg · K. Abram
Faurecia Emissions Control Technologies, Augsburg, Germany
e-mail: dennis.boennen@faurecia.com

H. Steinkilberg
e-mail: hannes.steinkilberg@faurecia.com

K. Abram
e-mail: kwin.abram@gmail.com

6.1   Introduction

Noise and vibration engineering of exhaust systems is continuously facing new challenges to provide solutions to requirements in terms of integration of new functionalities, products manufacturing robustness, dynamic fatigue and durability contribution to vehicles lightweight objectives. At the same time the pressure drop of the system within a given volume with tailored sound quality to specific needs of the application vehicles needs to be contained.

These challenges are eased by continuous innovations in products design and manufacturing process as well as system integration. In the past years, the leading exhaust system manufacturers have achieved weight reductions of more than 25 % of the system, knowing that the exhaust system weight represents only 2 % of the total vehicle weight. The lightweight structures today integrates—in standard engineering practice at Faurecia—the sound absorption materials, the materials thickness reduction, hydro-forming tubes, joining technologies, multi-layered/constrained layered materials as well as passive, reactive and active control valves. The employment of these technologies with an in-depth understanding of noise and vibration generating mechanisms and systems behaviors opens new potentials in lightweight structures over the next coming years.

An additional trend within the exhaust system development, especially due to downsizing of engine, is the introduction of active controlled components, such as loudspeakers, adjacent the exhaust line with the primary aim of changing the sound signature of the vehicle. Such systems could potentially be used for lightweight design. But here, the optimum target between sound enhancements, backpressure, and weight need to be addressed depending on the customer requirements.

6.2   Main Contributors to Lightweight Design

Background: The flow pulsations as well as vibration excitations are contained within exhaust systems by appropriate design of silencers architectures and systems integrations to an optimal system volume, pressure drops, sound transmission loss, vibration insulations and costs. In lightweight exhausts—in addition to the tailpipe noise— the sound radiation of components and modules play a significant role in the total sound quality of the system. The parameters affecting the vibro-acoustic behavior of the system are the mass, dynamic stiffness and the damping factors as well as the radiation efficiency of the structure in its environment. The use of lightweight materials such as aluminum, titanium, magnesium alloys or even composite materials for mass reduction have been investigated which has shown to have limited applications only in cold-end of exhaust line products due to the temperature limitations. Furthermore, the ratio of material Young modulus over the material density as well as temperature limitations and increase of costs versus steal alloys, limits the justification of employing such materials. The absorption materials have

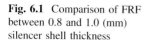
Fig. 6.1 Comparison of FRF between 0.8 and 1.0 (mm) silencer shell thickness

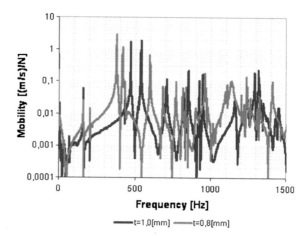

largely contributed not only to reduce the cavity acoustics as well as high frequency sound pressure reductions for a better sound quality of the silencers but also to absorb excitations to the housing and thereby to reduce the sound radiation of the parts.

Lightweight structures: The conventional approach to increase the dynamic stiffness of the structures will allow increasing the first eigenfrequencies to less influenced frequency ranges by the source of excitations. This could be achieved by an appropriate design of the shells (i.e. hallow vs. plates) of the boundary conditions (i.e. baffle positions within the silencers) and implementing stiffening deformations, ribs and dimples on specific positions of considerations. Particular attention should be given to not increase the frequency response of the system within audible bands where the combined radiation efficiency and acoustic weighting functions have the maximum levels. The following mobility function—measured as direct Frequency Response Function (FRF) in cold condition—shows clearly the increase in eigenfrequency of the modes and a slight reduction in the mobility level of the first mode as well as the overall levels in comparison of 0.8–1.0 (mm) shell thickness. (Figure 6.1).

The next two colormaps—shown in Fig. 6.2—are the sound pressure level measurements at 100 mm from the surface on an engine test bench in run-up conditions. It can be clearly observed that the response frequencies have shifted to a higher frequency ranges with also lower levels in higher frequency bands going from 0.8 to 1.0 (mm) shell thickness.

This trend is also confirmed by the variation analysis of parameters using shell's analytical model. The mobility is defined as the response velocity over the dynamic excitation in the following equation [1].

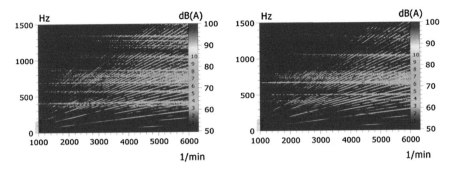

Fig. 6.2 Colormap plot of the Sound Pressure Level (SPL) at 100 (mm) from the center surface of a silencer with 0.8 (mm) –left– compared to 1.0 (mm) –right– shell thickness in run-up engine (6 cylinder) test

$$\frac{v(R_l, R_c)}{F(E_l, E_c)} = \frac{i \cdot \omega}{M} \cdot \sum_{m=1}^{\infty} \sum_{n=1}^{\infty} \left(\frac{\frac{1}{\omega_{mn}^2 - \omega^2} \cdot [f(m, R_l) \cdot f(n, R_c)] \cdot}{[f(m, E_l) \cdot f(n, E_c)]} \right) \qquad (6.1)$$

where:

$$\omega_{mn} = \tilde{\omega}_{mn} \cdot \left(1 + i \cdot \frac{\eta}{2}\right) \qquad (6.2)$$

and:

- v/F is the mobility function
- m and n are the longitudinal and circumferential mode numbers
- E and R are the excitation and response positions in longitudinal and circumferential locations respectively according to the index l or c
- M is the modal mass
- η is the loss factor
- ω_{mn} is the damped eigenfrequencies

The eigenfrequencies of the shell are calculated using Timoshenko-Love model. Then the radiated Sound Power can be calculated using the relation between surface vibration velocity and sound power in the following equation:

$$SWL_R = \sigma_{rad} \cdot A \cdot \rho_0 \cdot c_0 \cdot \left\langle \left| \frac{v}{F} \right|^2 \right\rangle \qquad (6.3)$$

where:

- SWL_R is the normalized sound pressure level to dynamic loads
- σ_{rad} is the Radiation Efficiency model
- A is the radiated surface area
- ρ_0 is the air density
- c_0 is the sound celerity in air

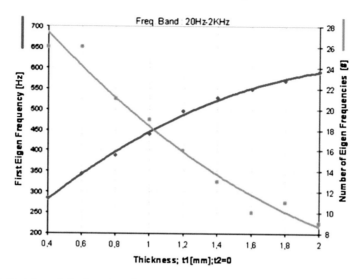

Fig. 6.3 Variation of the first eigen frequency as well as the mode density (number of modes) versus the shell thickness

The calculation of the radiated Sound Power Level (SWL_R) is performed using Leppington et al. average radiation efficiency model [2]. Furthermore, the sound pressure levels—normalized to unit dynamic load excitations—can be calculated by converting the mobility and using the average radiation efficiency models.

The parametric analysis of the thickness in the following Fig. 6.3 shows that the first eigenfrequency increases and the number of modes representing the modal density decreases as the thickness increases within the frequency band of analysis.

The vibration propagations could also be contained through impedance changes between parts/interfaces. The impedance changes are achieved in general by local modifications (i.e. point contacts) in mass/stiffness which contains the wave propagations through parts.

The increase of stiffness is also extensively used as a first approach to increase the first eigenfrequencies to higher frequency bands where the excitation—and in particularly the engine major harmonics—have lower levels. The in-depth knowledge of the local structural dynamic responses to excitations in combination with numerical topology optimizations would allow significant increase in eigenfrequencies under considerations.

The structural damping—in its general terminology—is the most promising parameter which gives additional perspective in noise reduction of lightweight structures.

In its material content, the vibration energy is dissipated within the material due to its loss factors and thereby the level of radiated noise could be reduced consequently. However the loss factor of the metallic material remains extremely low even using special treatments (i.e. tempering, coating, etc.), making the composite materials a potential candidate for this application by appropriate design of the system in order to keep this material away from the heat sources.

Fig. 6.4 Damping factors of different silencers as a function of frequencies

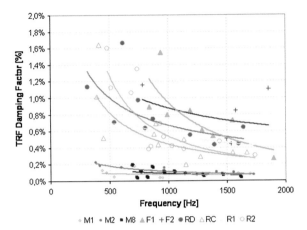

Furthermore, appropriate design of structural deformations and dimples, not only contributes to the dynamic stiffness but also in some extent to the structural damping of the part as well as reducing the radiation efficiencies. [3].

The following graphs in Fig. 6.4 are the measurement results of damping obtained by using log-decrement-method evaluation on Time Response Function for each resonance—as well as its Power fittings curves.

From Fig. 6.4, it can be observed that the damping factor of the lower frequencies are higher and, therefore, suggesting that the first natural modes of the system is more easily damped due to the dissipation of vibration energy in interfaces and in damping materials. A power-curve-fitting plot is made for each silencer to highlight this behaviour.

The damping factors for Silencer M1, M2 and M8 are extremely low due to its simple metal shell definitions and without any insulation materials or damping interfaces. The Silencers F1 and F2 are smaller silencers in terms of dimensions and therefore have higher eigenfrequencies but also exhibits higher damping factors.

The damping could also be achieved by friction losses between the interfaces, known as coulomb damping. An application of this behaviour is the multi-layer skin materials employed in particularly for rolled silencers. This design significantly reduces the response levels by the mechanism of the friction damping between the layers. However, due to the decreasing inertia, which decreases the stiffness, the eigenfrequencies shifts in lower frequency bands which could be subjected to higher excitation levels. Additionally, particular care should be given to the manufacturing process in order to assure the surface contacts between the layers throughout the life cycle of silencers.

The following graphs in Fig. 6.5 shows the comparison between single and multi-layers skin silencer showing clearly the reduction of the response levels as well as the eigenfrequencies of the shell. These results can also be confirmed when performing engine bench tests.

Furthermore, different design parameters are considered to efficiently damp the vibration of the structure at the interfaces and therefore reduce the radiated noise within the frequency of interests.

Fig. 6.5 Comparison of FRF between single-layer with 0.8 (mm) silencer shell thickness and double-layers with 0.5 and 0.3 (mm) thicknesses of the silencer shell

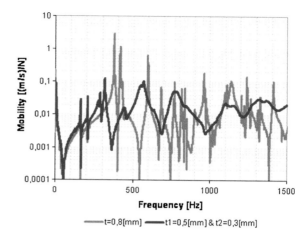

6.3   Acoustic Exhaust Valves—an Exhaust Mass Reduction Enabler

Background: Using valves within an exhaust system to improve acoustic performance is a tool which has been used for decades in a wide variety of applications. It is quite intuitive to most automotive acousticians that the addition of backpressure to an exhaust system should result in a reduction of sound radiating from the tailpipe. It's also intuitive that using a valve to by-pass the mufflers in an exhaust system will yield an increase in sound radiating from the tailpipe. However, the use of valves as an exhaust system mass reduction enabler is not as well understood and is the focus of this section.

There are many different acoustic valve designs which generally fall into combinations of these 3 categories:

1. **Throttling or By-Pass**: For throttling valves, 100 % of the gas in an exhaust system is forced to travel through the valve. Throttling valves never completely close (otherwise the engine couldn't continue to run) but can provide up to 95 % area restriction in some cases. By-Pass valves, on the other hand, typically attempt to have 100 % coverage in their closed position forcing all of the gas through an alternate path (by-pass path).

2. **Active or Passive**: Active valves have an actuator (vacuum, electric, etc.) and are opened/closed depending upon the engine operating condition and control logic. Most active valves operate in a 2-position fashion: open or closed. However, in some cases they can be continuously variable. Passive valves are typically driven by an exhaust gas parameter (backpressure, mass flow, temperature, etc.) and do not have a controller. The most common passive valve is one which is held in a "closed" position by a spring and then opens as the backpressure force on the valve exceeds the spring force which causes the valve to open. However, passive valves typically do not completely open since the

spring force is always being balanced by the backpressure against the valve. This also means that passive valves are continuously variable and can operate at any position between maximum closed and maximum open depending upon the engine operating condition.

3. **2-position or Continuously Variable**: Typically, active valves are 2-position and passive valves are continuously variable. However, there are exceptions to this and the benefits/penalties of this design choice are quite extensive. Therefore, this choice must be considered as a third design choice category.

Mass Reduction Opportunities Using Exhaust Valves: When valves are used within an exhaust system to reduce mass, the objective is to implement a valve in a manner which enables the reduction of both the muffler size and complexity (more complex mufflers have added mass due to additional internal components). Each of the 3 categories of design choices will affect the exhaust sound in a different manner and therefore will have a different impact on mass reduction. The following is a summary of the impact of each choice on mass reduction.

Throttling Valves—Active Versus Passive and 2-Position Versus Continuously Variable: Throttling Valves generally offer improved mass reduction opportunities over By-Pass Valves due to their ability to suppress very low frequencies (all the way down to 0 Hz in theory). This is accomplished due to the pressure drop across the valve (typically 2–10 kPa) which introduces an impedance to the traveling acoustic wave. As the wave attempts to travel through the valve, its velocity (which is superimposed on the exhaust mean particle velocity) is accelerated through the valve orifice with high velocity and turbulence which converts the wave energy into heat. The result is a reduced level wave downstream of the valve. The reduction is the same for all frequencies of the same acoustic pressure amplitude. However, since lower frequencies tend to have higher amplitudes, lower frequencies typically have greater sound reductions. This type of broad band acoustic attenuation typically reduces exhaust sound by 5–10 dB. An example of measured exhaust system sound on an SUV test vehicle with and without a throttling valve is shown in Fig. 6.6 (compare the black and red lines). Notice how all frequencies are attenuated at a similar level with the red line but the attenuation is greatest at the lowest frequencies. These throttling results are quite impressive but there is a second benefit of throttling valves compared to by-pass valves which provides even more interesting results.

Since throttling valves can be used to convert traveling wave energy into heat, they can also be used as an acoustic damper to suppress acoustic resonances within an exhaust system. By suppressing (or damping) the resonances using a throttling valve, it is possible to reduce muffler volume. For a throttling valve to be effective at suppressing acoustic resonances, the valve must be located at a velocity anti-node. An example of a velocity anti-node would be the inlet or outlet ends of a pipe which is experiencing an acoustic standing wave resonance (a half-wave resonance for example). In a second example on the SUV test vehicle, a throttling valve was applied to a pipe with a resonance at 200 Hz which created an undesirable rise in sound as shown in Fig. 6.7 (compare the black line to the blue line).

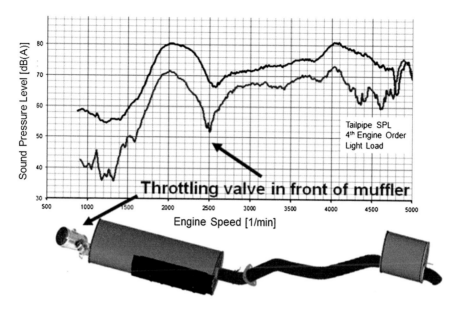

Fig. 6.6 Impact on throttling valve in front of muffler—vehicle measurements; impact on 4th order

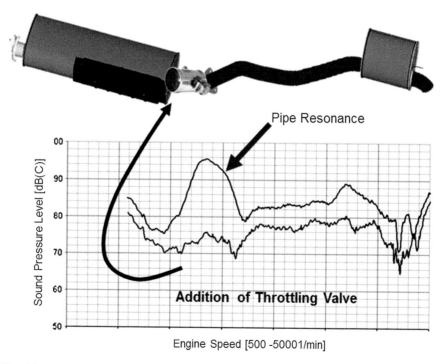

Fig. 6.7 Impact of positioning of throttling valve—vehicle measurements; impact on 4th order

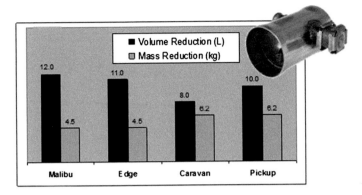

Fig. 6.8 Example of volume and mass saving using a throttling valve

In this example, the throttling valve effectively removed the resonant behavior creating an acoustically linear response which is much more desirable in most applications. Resonance suppression can reduce sound pressures by up to 20 dB (as shown in this example) at the peak resonant frequency.

Since both of the attenuation effects (broad band suppression and resonance damping) can occur simultaneously with an optimal application of a throttling valve, muffler volumes can be reduced by 25–35 % using this technology as shown in Fig. 6.8.

This volume reduction can typically yield mass reductions of 3–5 kg due to the smaller mufflers and simplified muffler internals. But this benefit does not come without a penalty which is the addition of backpressure. If the throttling valve remained in a closed position at high mass flow rates the backpressure would be excessively high. To counter this negative backpressure effect, throttling valves must open as mass flow is increased. The red line in Fig. 6.9 shows the measured backpressure of a standard exhaust system without the use of a valve. The system

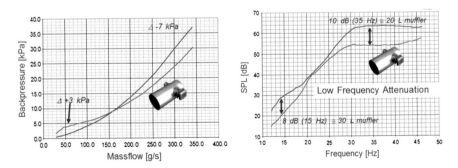

Fig. 6.9 Impact of throttling valve in backpressure; *red line*—traditional system; *green line*—system with throttling valve (*left*); Impact of throttling valve low frequency attenuation; *red line*—traditional system; *green line*—system with throttling valve (*right*)

was then modified by implementing a passive throttling valve (which increased backpressure at low mass flow conditions) and also modified by reducing the backpressure of the mufflers in the exhaust system. The resulting backpressure shown in the green line for Fig. 6.9 (left) has increased backpressure below 150 g/s of flow and then transitions to having decreased backpressure above 150 g/s. The low frequency acoustic data acquired on the vehicle for these two exhaust systems below 150 g/s is shown in Fig. 6.9 (right). The 8–10 dB reductions resulting from the added backpressure would have required 20–30 L of additional muffler volume without the use of a valve due to the extremely low frequency energy (below 50 Hz).

Passive Throttling Valves use a spring and an offset shaft which allows the valve to open as the mass flow (and backpressure) is increased to reduce backpressure at high mass flow operating conditions. By choosing an appropriate spring rate and pre-load, the valve will provide optimal acoustic attenuation at low mass flow rates while providing acceptable backpressure at high mass flow rates. Passive Throttling valves are also continuously variable which enables smooth transitions throughout the engine operating conditions yielding subjectively desirable transitions in sound.

Active Throttling Valves can provide all of the same benefits as a passive throttling valve with the option for additional control of the sound to further improve the sound experience for the vehicle occupants. However, this optimal performance can only be accomplished if the active valve is continuously variable which is quite uncommon due to the complexity of implementation. Therefore, most active throttling valves are 2-positon which greatly limits their ability to be optimized for mass reduction. 2-position active valves are very popular for high performance vehicles by allowing the exhaust sound levels and exhaust back-pressures to coincide with two specific driving conditions: Sporty and Quiet. Attempting to separate all driving conditions into two categories of sporty or quiet is certainly not an optimal approach but has been proven over many years to create an acceptable driver experience and continues to be used by many vehicles today. When a sporty driving condition is encountered, the valves are commanded open to yield maximum power and maximum sporty sound. However, during non-sporty driving (cruising on the highway for example), the active valves are commanded to be closed to create a quiet driving environment. The mass reduction opportunities for 2-position active throttling valves is highly dependent upon the trade-offs between quiet, sporty, engine power and control logic. However, up to 25 % mass reduction is possible.

By-Pass Valves—Active Versus Passive and 2-Position Versus Continuously Variable: One of the key issues with rear muffler layouts is an effect commonly referred to as the tailpipe-to-muffler Helmholtz resonance. This lumped parameter resonance occurs at a natural frequency where the gas in the tailpipe acts as a mass and the gas in the muffler acts as a spring creating a classical resonance (Resonance Frequency = (Spring rate/mass)^0.5). It is a common issue for rear muffler designs that this resonance will coincide with engine firing frequency at idle or at low operating speeds which creates undesirable tailpipe sound. By implementing a by-pass valve on the rear muffler/tailpipe configuration, the tailpipe-to-muffler

Helmholtz frequency can be shifted to a more acceptable frequency. Also, since these by-pass valves tend to be 100 % closed at low mass flow rates, the closed backpressure and acoustic transmission loss can be increased significantly at low mass flow rates.

Compared to throttling valves, by-pass valves are not effective for resonance suppression and minimally effective for broadband attenuation. However, they are often lower cost and/or more simple to implement. With optimization, 15–25 % reduction in muffler volume is possible with typical mass reductions of 5–10 %.

Passive By-Pass Valves are typically installed inside a muffler which is mounted at the rear of the vehicle to minimize the effect of high temperature gas on the valve spring. At low mass flow rates these valves are closed which creates a muffler configuration with increased acoustic attenuation relative to the same muffler with the valve open (typically 5 dB of additional attenuation at low engine speeds). While passive by-pass valves have a well-defined closed position, the open position is continuously variable which creates smooth acoustic transitions between varying mass flow conditions.

Active By-Pass Valves are typically installed in a tailpipe of an exhaust system with 2 tailpipes. It is important that these valves have a 2-Position design because there needs to be either no flow through the closed valve for optimal acoustic attenuation or completely unrestricted flow through the open valve to prevent unacceptable flow noise. If an active by-pass valve would be continuously variable, the flow noise created during partial opening events would not be acceptable in a tailpipe application. Much like the 2-position active throttling valve, the 2 position active by-pass valve is well matched for sporty applications where defining all driving conditions as either quiet (valve closed) or sporty (valve open) has proven to be quite acceptable.

Summary: The table in Fig. 6.10 shows a ranking of the 5 most popular design category combinations. In order to maximize the opportunities for mass reduction, the acoustic performance for each is ranked from best to worst.

Acoustic/Mass Performance Ranking	Complexity Ranking	Throttling or By-Pass	Active or Passive	Continuously Variable or 2-Position
#1 (Best)	#5	Throttling	Active	Continuously Variable
#2	#2	Throttling	Passive	Continuously Variable
#3	#4	Throttling	Active	2-Position
#4	#3	By-Pass	Active	2-Position
#5	#1 Lowest	By-Pass	Passive	Continuously Variable

Fig. 6.10 Exhaust valve ranking based on mass reduction opportunities and complexity

Throttling valves outperform by-pass for mass reduction optimization and are ranked as the top 3 acoustic performers. The best overall acoustic performing is the active continuously variable throttling valve which is also the highest in complexity. However, the 2nd highest performer is the passive continuously variable throttling valve which also ranked 2nd in complexity. It should be noted that if the vehicle is extremely sensitive to backpressure and/or has specific sporty sound requirements, the 2-position Throttling Active Valve may be a more desirable choice over the Passive Continuously Variable Throttling Valve. Finally, the passive by-pass valve (typically located within the muffler) is the lowest performer but is also the lowest complexity. The combination of low complexity while still providing quite respectable acoustic attenuation has made this the most popular exhaust valve on the list with millions being sold over the past 2 decades.

In the end, the decision on which valve design to use will be based upon many factors such as sporty vehicle characteristics, marketing strategy, performance requirements, mass, complexity and durability. Each OEM ultimately must create their own criteria for the benefits and penalties of each design in order to determine which solution is optimal for each vehicle platform.

6.4 Weight Reduction and Active Exhaust Systems—Possibilities and Constraints

Overview: Whenever active exhaust systems are mentioned in conjunction with lightweight, the objectives of applying these systems can be quite different. On the one hand, pure sound design systems are considered to be enabler for the use of downsized engines premium or sporty vehicles. On the other hand, a weight saving by making use of active noise cancelation has often been proclaimed. Within this chapter, possibilities as well as constraints of these different approaches are discussed so that positive and negative aspects can be weighted.

Currently, all active exhaust systems are based on the integration of one or two loudspeakers adjacent to the exhaust line. This means that for such an application a custom-made loudspeaker is required. It has to be heat and condensate resistant and requires a special acoustic tuning. Since there are certain limits to the temperature resistance of a speaker chassis—especially the rubber softroll and the gluing is critical—the speaker needs to be mounted in a side branch of the exhaust line. By optimizing the shape of the connection and tuning the length of the connection pipe, the resulting gas temperature in front of the speaker can be significantly reduced.

A control unit with integrated amplifier receives all relevant engine information over the CAN bus. The control unit computes the appropriate anti-noise or sound signal. This signal is amplified and passed to the speaker.

For cancellation, an additional error sensor in the tailpipe is required and the system is running in closed loop. A sophisticated algorithm drives the speaker to create a signal with the same amplitude, but with inverted phase to the original sound wave from the engine. For sound design only, the sensor is omitted and the system is running in open loop.

Other technologies, such as oscillating valves, have proven their feasibility for cancellation of especially low-frequency order noise. But due to their enormous mechanical complexity and their conceptually inevitable backpressure increase, these concepts have been abandoned.

Active Systems to Enable Downsizing for Premium Vehicles: Buying a car often is an emotional decision. Therefore, the sound of a car is an important criterion to communicate the brand image and to emphasize the character of the car. The sound of an entry level car should differ from the sound of a middle-class or premium class vehicle. And the sound signature of a sports car is closely linked to the subjective impression of the engine performance.

Because of the wide-spread use of turbochargers, 4-Cylinder engines are becoming more and more present in premium as well as sports cars, where these engines are replacing previous generations of 6-cylinder engines. In the motor press, these engines are getting excellent ratings for power and torque, but their sound quality is rather inadequate. Here, dynamic sound generation helps to give the car an individual, class- and brand- representing sound signature.

Adding an exhaust sound system to a vehicle means adding an extra weight of approx. 3.5 kg, whereas the engine downsizing effects a weight reduction of approx. 50 kg. Since active exhaust systems can be used to increase the sound perception of smaller engines as well as diesel engines in powerful cars, this technology can be considered a downsizing enabler—not for all cars, but for some prestigious vehicle models. In Fig. 6.11 an example is shown how the sound signature of an active sound system could look like. Here, the sound system is used to create an 8 cylinder sound signature from a 4 cylinder engine.

Whenever an OEM follows a more holistic approach, active exhaust systems could offer some contributions to weight saving efforts.

For a significant contribution to weight saving, an active exhaust system needs to reduce passive muffler volume by a significant amount. On the other hand, noise reduction by anti-noise means that the noise from the engine does not have to be greater than the SPL the speaker can generate. Otherwise the active system will not be able to attenuate the engine noise any more. This means that the reduction of passive muffler volume is limited by the acoustical power of the active system.

For efficient cancellation—to enable the reduction of passive muffler volume— one key element is the loudspeaker. There are certain aspects to consider. Since the speaker needs to be as powerful as possible, the question arises what the maximum acceptable size and power would be for usage in light vehicles. Due to packaging constraints, speakers larger than 8 inches will not be acceptable. For such speakers typical maximum power consumption ranges between 100 and 150 W RMS. A further improvement would not be an increase in power, but in speaker efficiency. Here, an important aspect is the magnet weight. Larger and more efficient loud-speakers will easily get really heavy—because of their ferrite magnets. With the focus on lightweight, the use of neodymium becomes mandatory—at least for

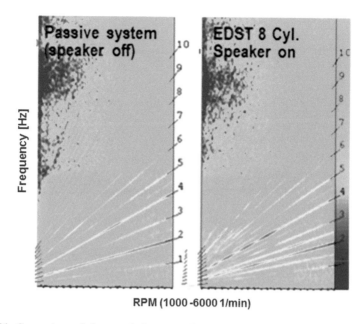

Fig. 6.11 Comparison of the sound signature of a 4 cylinder turbocharged vehicle with and without a sound system in a full-load run-up condition [4]

larger speakers. Weight difference between ferrite and neodymium can be as much as 1 kg.

Since relevant engine orders start at approx. 30 Hz, a sufficient speaker performance in the lower frequency range is required. In consequence, this means that the speaker chassis needs to be designed for a high membrane displacement.

For some cars—smaller gasoline and some diesel engines—passive mufflers will completely be replaced by an active system in combination with some small resonators. For more powerful engines, for example 4-cylinder turbocharged gasoline engines with up to 200 kW, there will still be the need for passive muffler volume—which then can be reduced for example from 30 to 12 l. This requires that cancellation is working well. In Fig. 6.12 the comparison of a possible system layout, which is currently implemented on the internal demonstration vehicle, can be seen. The shown volume reduction of the passive system correlates to a weight saving of 4–5 kg, without taking the additional weight of the speaker into account. That means that in this case an overall weight reduction of nearly 1 kg could be achieved.

In Fig. 6.13 the measured results obtained on the rollerbench under full-load condition is shown and compared to simulated data. The blue line represents the second order sound pressure level of the serial design layout. The change of the layout leads to a changed primary noise (black line)—here loudspeaker is not running. The possible emitted sound pressure level of the speaker is shown in red. That means in optimum conditions with a perfect working algorithm the residue

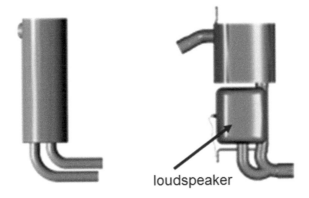

loudspeaker

Fig. 6.12 Serial design (*left*) in comparison to active exhaust layout (*right*)

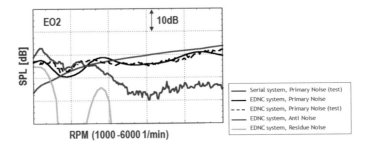

Fig. 6.13 Results of rollerbench measurements of serial design exhaust (*blue line*) compared with measured and simulated data (*black lines*) of the active exhaust system and the possible residual noise (*green curve*) when the loudspeaker giving a certain output (*red curve*)

noise represented by the green curve could be reached. It can be seen that for all frequencies where the anti-noise curve is above the curve of primary noise, a full cancellation can be achieved.

Obviously, such reduction cannot be achieved in real conditions. When the decision for an active exhaust system has been made—usually driven by desires for better acoustics—these systems can also be tuned for lightweight. But this means that some prioritization has to be made: a decision, if the system should be optimized for tailpipe noise, backpressure or weight reduction.

6.5   Conclusion

Within this paper the aspect of traditional lightweight as well as the implementation of new component as lightweight enabler in the exhaust system has been discussed.

- The structural dynamics of an exhaust system can be largely influenced by changing the dynamic stiffness in order to shift the eigenfrequencies in the desired regions. This allows dissociating the structural responses from the excitations and therefore to reduce the radiated sound pressure level. A thorough understanding of the system behaviour is mandatory.
- As damping brings dissipation of vibration energy within the structure, it opens additional opportunities for noise containments. Adding damping at interfaces, such as baffle/skin connection is one of the possibilities as well as using multilayer structures or new materials with increased structural damping.
- Adding passive or active valves to the exhaust system could have huge benefit with regard to weight saving, but an application of valve has to be done with care, otherwise negative functional impacts such as backpressure increase could occur.
- Exhaust Dynamic Sound Generation enables to increase the sound performance of the vehicle. The direct impact on the exhaust system weight is negative, but looking at the overall weight of a downsized vehicle the weight balance is positive.
- Exhaust Dynamic Noise Cancellation systems are used in order to reduce unwanted noise. Here a compromise between lightweight, acoustics, backpressure has to be found.

References

1. Wallace CE (1970) Radiation resistance of a rectangular panel, J Acoust Soc Am—20 July 1970 Arizona State University, Tempe, Arizona 85281
2. Leppington FG, Broadbent EG, Heron KH (1982) The acoustic radiation efficiency of rectangular panels. Proc R Soc Lond Math Phys Sci 382(1783):245–271
3. Binder B (2013) Untersuchung von geometrisch versteiften Mantelkonzepten für Pkw-Schalldämpfer in Leichtbauweise, Master's Thesis, Hochschule Augsburg—University of Applied Science at Faurecia Emissions Control Technologies Germany GmbH—March 2013
4. Bönnen D, Kim H-J, Steinkilberg H, Zintel G (2013) Development methodology for active exhaust systems, DAGA 2013, Merano

Chapter 7
A Patch Transfer Function Approach for Combined Computational-Experimental Analysis of Vibro-Porous-Acoustic Problems

Jan Rejlek, Eugenius Nijman, Giorgio Veronesi and Christopher Albert

Abstract Driven by both the ever-increasing tightening of legal regulations and the growing customers' expectations, the noise, vibration and harshness (NVH) is becoming a crucial aspect in the vehicle development process. To achieve the NVH targets set for modern vehicles, sound insulation materials became an indispensable instrument to improve the vibro-acoustic behaviour. Typically, the sound insulation materials take advantage of so-called porous materials, which exhibit favourable properties when it comes to structural damping as well as transmission and absorption of sound. However, due to the highly complex material micro-structure and the sound propagation mechanisms involved the computational modelling of porous materials is a fairly challenging topic. An efficient yet accurate prediction of the NVH attributes of sound insulation materials therefore remains an unresolved issue. This chapter reports on recent developments based on so-called Patch Transfer Function (PTF) approach. Here the PTF approach is adopted for the analysis of coupled vibro-acoustic problems involving porous domains. The PTF is a sub-structuring technique that allows for coupling different sub-systems via impedance relations determined at their common interfaces. The coupling surfaces

J. Rejlek (✉) · E. Nijman · G. Veronesi · C. Albert
Virtual Vehicle Research Center, Area NVH & Friction, Inffeldgasse, 21/A, 8010 Graz, Austria
e-mail: jan.rejlek@v2c2.at

E. Nijman
e-mail: eugene.nijman@v2c2.at

G. Veronesi
e-mail: giorgio.veronesi@v2c2.at

C.G. Albert
Institut Für Theoretische Physik—Computational Physics, Technische Universität Graz, Petersgasse, 16, 8010 Graz, Austria
e-mail: albert@tugraz.at

© The Author(s) 2016
A. Fuchs et al. (eds.), *Automotive NVH Technology*,
Automotive Engineering: Simulation and Validation Methods,
DOI 10.1007/978-3-319-24055-8_7

99

are discretised into elementary areas called patches. Since the impedance relations can be determined in either numerical or experimental manner, the PTF approach offers very high degree of versatility and is hence well-suited for combining test and simulation data into one workflow. Efficiency of the methodology proposed has been demonstrated by means of a validation example consisting of a rigid cavity backed by a dynamic plate with porous treatment. The full-system measurements are compared with the PTF predictions based on component measurements and/or simulations.

Keywords Vibro-acoustics · Sub-structuring · Impedance coupling · Patch transfer functions · Porous materials · PU-probe

7.1 Introduction

Sound insulation materials are widely applied as dissipative treatments in vibro-acoustic systems. Whenever a vibrating structure radiates sound into an acoustic fluid, the insertion of sound insulation materials has three main effects: (i) structural loading and damping, (ii) decoupling the acoustic fluid from the structure (mass-spring systems) and (iii) adding absorption to the acoustic fluid. The sound insulation components are typically assembled by two or more material layers, from which so-called fluid-saturated porous materials constitute a substantial part. The porous materials consist of two phases—the solid one, which forms the skeleton, and the interstitial fluid phase, which is contained within the pores formed by the solid phase. Since both the transversal and longitudinal waves can exist in an isotropic solid, and since a longitudinal wave occurs in a fluid, three types of waves can propagate through a porous domain. As the frame and the fluid exhibit a strong mutual interaction, visco-thermal dissipation mechanisms take place.

Over the last decades, various mathematical models ranging from simple concepts to sophisticated methods [3] have been developed to represent the vibro-acoustic behaviour of sound insulation materials. In traditional numerical schemes [14], the influence on the structure is usually described by additional mass and damping. The damping of acoustic fluid is captured by solving a system with impedance boundary conditions. Required parameters may be estimated by material models, ranging from simple equivalent mechanical systems to phenomenological impedance models [10].

The current state-of-the-art in the numerical modelling of the vibro-acoustic behaviour of porous materials is represented by material micro-model based on the Biot theory [7], which is implemented in a finite element method (FEM) [2] or in a reduced transfer matrix scheme [1]. Although this approach allows for highly detailed material description, its practical implementation leads to very large computational burden, which limit its practical application for low frequency range. This becomes even more pronounced as far as industry-sized problems are considered. Moreover, a proper estimation of the material parameters required by the Biot model is not at all straightforward and is hence the reason, why are these parameters often not available in practice.

Similar to numerical approaches described above, different experimental techniques have been developed to characterise the vibro-acoustic properties of sound insulation materials. Mechanical parameters of sound insulation materials may be obtained in dynamic stiffness tests and impedances are measured in a standing wave tube or in-situ on the material surface [15]. An impedance tube is typically used for estimation of the normal impedance, which allows then the derivation of sound absorption coefficient and transmission loss [12]. Reverberation chambers and transmission suites are used for the measurement of, respectively, sound absorption coefficient and transmission loss under diffuse sound field conditions according to [13]. As these testing procedures involve large, special-purpose environments, alternative, non-standardised measurement procedures based on small reverberation cabins [6] have been developed over the past years. In order to determine the dynamic properties of damping layers, Oberst test method can be applied [17].

As far as Biot model is considered, a set of nine material parameters must be experimentally determined in order to provide the numerical model with corresponding data. Hence, highly dedicated, laboratory apparatuses need to be utilised in order to assess all required physical quantities [3].

In this chapter the Patch Transfer Function (PTF) approach is applied for the analysis of a coupled vibro-porous-acoustic problem. Although the PTF method has been originally proposed by Ouisse et al. [18] for the analysis of acoustic problems, its application field further rapidly expanded towards more general problems. Pavic et al. [19] used the PTF technique for source characterisation, Maxit et al. [16] improved the approach for structures coupled to heavy fluid, Guyader et al. [11] and Chazot et al. [9] solved a transmission loss problem and Aucejo et al. [4] have conducted a study on the convergence of the methodology when applied to heavy fluids. Recently, Rejlek et al. [20] and Veronesi et al. [21] have shown the implementation of the PTF method for coupled vibro-acoustic problems based on hybrid component testing and simulation.

The remainder of the chapter is organised as follows. Section 7.2 provides an overview of the underlying mathematical formulations behind the vibro-porous-acoustic problems considered in this article. Next, the basic principles behind the PTF methodology are presented in Sect. 7.3. In the Sect. 7.4 a novel experimental trim characterisation procedure is proposed and validated. Section 7.5 reports on the application of PTF approach on a coupled vibro-porous-acoustic problem and further presents the results of a validation study conducted on full-system. Finally, the results obtained are discussed in conclusions and an outlook for the ongoing and future activities on this topic is given.

7.2   Problem Definition

Consider a three-dimensional fully coupled vibro-acoustic problem, see Fig. 7.1. The structural part Ω_S consists of a thin flat plate subjected to (i) boundary conditions imposed at the physical boundary Γ_{sa}, Γ_{sp}, (ii) a harmonic point force $Fe^{j\omega t}$ acting in the out-of-plane direction at the position r'_F and (iii) the acoustic pressure loading p arising from the coupling to an acoustic domain along the structural-acoustic interface Γ_{sa}, or to the pressure loading and velocity arising from the coupling to a porous domain at the interface Γ_{sp}.

The porous part Ω_p consists of an isotropic porous material applied on a partition of the structural domain at the interface Γ_{sp}. The skeleton of the porous domain is subjected to (i) normal and tangential velocities and stresses due to the coupling to the structure at the interface Γ_{sp} and (ii) the pressure and normal velocity given by the interaction with the acoustic domain at the interface Γ_{pa}. The interstitial fluid in the porous domain is subjected to (i) normal velocities and normal stress due to the coupling to the structure at the interface Γ_{sp} and (ii) the pressure and normal velocity given by the interaction with the acoustic domain at the interface Γ_{pa}. The porosity influences the coupling at the porous-acoustic interface.

The acoustic part Ω_a consists of a closed physical boundary Γ_a filled by an acoustic fluid. The acoustic part is subjected (i) to boundary conditions defined at Γ_a and (ii) the normal velocity and pressure distributions at the structural-acoustic interface Γ_{sa} and at the porous-acoustic interface Γ_{pa} accounting for, respectively, the coupling effects due to the plate dynamics and the coupling to the porous domain. The Helmholtz equation is governing the steady-state acoustic pressure perturbation field p(r) in the acoustic domain.

Fig. 7.1 A coupled vibro-porous-acoustic problem—$\boldsymbol{\Omega}_s$ structural part, $\boldsymbol{\Omega}_p$ porous material, $\boldsymbol{\Omega}_a$ acoustic fluid domain, $\boldsymbol{\Gamma}_\bullet$ the corresponding interfaces and $\boldsymbol{F}_\bullet$ the loads

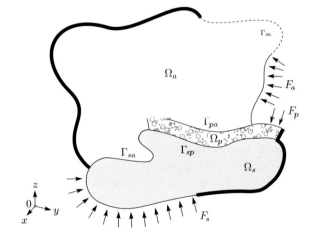

The dynamic excitations show a time-harmonic behaviour $e^{j\omega t}$ with a circular frequency ω. The corresponding time-harmonic response of the structural-acoustic system is thus described in terms of steady-state displacement w and the acoustic pressure field p.

7.3 Basic Concepts of the PTF Approach

The PTF is a sub-structuring method. It consists of partitioning of the continuous system into sub-systems and coupling them at their common interfaces via impedance or mobility relations. Each interface is subdivided into N elementary surfaces ∂S called patches. For practical reasons the interfaces correspond to physical interfaces formed between the respective sub-domains. In order to couple different sub-domains a set of transfer function called Patch Transfer Functions (PTF) must be determined. These transfer functions are based on dynamic field quantities spatially averaged over the patches. Once the PTFs are determined the coupling procedure consists of the application of a superposition principle [8].

Let us consider an interface S between two sub-domains, and let us divide it into small elementary surfaces ∂S_j, see Fig. 7.2. For the jth patch centred at position r of the mth sub-system the averaged pressure and velocity are given by

$$\bar{p}_j^m = \frac{1}{\partial S_j} \int_{\partial S_j} p_j^m(\mathbf{r})dS \qquad (7.1)$$

$$\bar{v}_j^m = \frac{1}{\partial S_j} \int_{\partial S_j} v_j^m(\mathbf{r})dS. \qquad (7.2)$$

The transfer function between an excited patch i and a receiving patch j defined on the sub-system m can be expressed in terms of impedance by

$$Z_{ij}^m = \left.\frac{\bar{p}_i^m}{\bar{v}_j^m}\right|_{\bar{v}_{k\neq j}^m=0}, \qquad (7.3)$$

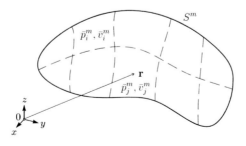

Fig. 7.2 Discretisation of the interfaces into patches and assessment of averaged quantities

and in terms of mobility by

$$Y_{ij}^m = \left.\frac{\bar{v}_i^m}{\bar{p}_j^m}\right|_{\bar{p}_{k\neq j}^m=0}.$$ (7.4)

   To obtain Z_{ij}, we excite with a surface velocity $\bar{v}_j$ at a certain patch j and block all other patches so that $\bar{v}_{k\neq j} = 0$. Then the ith row of Z is given by the ratio of all surface pressures to the single velocity, see Eq. (7.3). To determine Y_{ij}, we excite with a surface pressure $\bar{p}_j$ at patch j and leave all the other patches free with $\bar{p}_{k\neq j} = 0$. The ith row of Y is given by the ratio of all surface velocities to the single pressure, see Eq. (7.4). The first case will be the natural method for acoustic and porous domains, while the latter for structural ones. In the numerical model, we can just successively apply these boundary conditions to obtain the blocked pressures or free velocities on the surface. In principle this can be also done experimentally, even if the implementation can be difficult.
   Determination of the impedance and/or mobility matrices is not only restricted to an approach described above, but can be also derived in an indirect manner, which is one of the ongoing research activities.
   In experimental setup or numerical analysis the patch transfer functions can be obtained in a straightforward way by splitting the interface into N patches and averaging the quantities over each patch. The resulting patch values can be obtained by averaging or integrating over measured or simulated quantities inside a patch area. In order to avoid spatial aliasing the sampling spacing has to obey to the Nyquist criterion accounting for the corresponding wavenumber. In order to decide upon the patch discretisation, again the Nyquist sampling criterion has to be applied yielding the appropriate patch dimension $d \leq \pi/k(f_{max})$ [22]. This rule ensures that the sound radiated from the patch discretisation is correctly accounted for.

7.3.1 Coupling Procedure

A superposition principle is used to couple the systems at their common interface. The coupled response of one sub-system is given by a sum of the two virtual configurations. In case of a fluid sub-system the two configurations are (1) the pressure response due to internal sources if the coupled interface is blocked and (ii) the pressure response due to the yet unknown interface velocity of the coupled system, when the internal sources are switched off. In case of a structural sub-system the two configurations are (i) the velocity response due to internal sources if the coupled interface is free and (ii) the velocity response due to the yet unknown interface pressure of the coupled system, when the internal sources are switched off.

7.3.2 Coupling a Structural Domain with Two Fluid Layers

Let us consider the coupled system in Fig. 7.3 where a thin steel plate loaded by a point force F is coupled to a fluid layer representing trim, which then couples to an acoustic rigid cavity. The fluid layer is in fact an equivalent fluid used to represent the porous domain. The continuity conditions for pressures and velocities at the interface between the structural and the porous domains are

$$p_1 = p_1^S = p_1^T$$
$$v_1 = v_1^S = v_1^T,$$
(7.5)

where the superscript S denotes the structure and T the trim. Between the porous and the acoustic domain following expressions hold

$$p_2 = p_2^T = p_2^C$$
$$v_2 = -v_2^T = v_2^C,$$
(7.6)

where the superscript C denotes the cavity. The pressure and velocity relation between the structural and acoustic domains read

$$p_1 = p_1^S = p_1^C$$
$$v_1 = v_1^S = v_1^C.$$
(7.7)

The impedance relations at the structure-porous interface write

$$\mathbf{v}_1 = \mathbf{Y}^S \mathbf{p}_1 + \mathbf{y}^F F$$
(7.8)

$$\mathbf{p}_1 = \mathbf{Z}_1^T \mathbf{v}_1 - \mathbf{Z}_{12}^T \mathbf{v}_2,$$
(7.9)

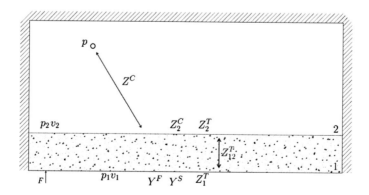

Fig. 7.3 A fully coupled vibro-porous-acoustic problem: a steel plate is coupled to a porous layer at interface 1 and successively to a fluid cavity at interface 2

where $\mathbf{Y}^S$ is the mobility matrix of the structure, $\mathbf{y}^F$ is a vector of structural transfer function between the point where the load is applied and all patches on the plate. $\mathbf{Z}_1^T$ and $\mathbf{Z}_{12}^T$ are, respectively, the surface trim impedance at interface 1 and the cross trim impedance between interface 1 and 2.

The impedance relations at the acoustic-porous interface write

$$\mathbf{p}_2 = \mathbf{Z}_{21}^T \mathbf{v}_1 - \mathbf{Z}_2^T \mathbf{v}_2 = \mathbf{Z}_2^C \mathbf{v}_2, \tag{7.10}$$

where $\mathbf{Z}_2^T$ and $\mathbf{Z}_2^C$ are the surface trim impedance and the surface cavity impedance at the interface 2 respectively. The fully coupled problem can be described in a matrix formulation:

$$\begin{pmatrix} -\mathbf{Y}^S & -\mathbf{I} & 0 \\ \mathbf{I} & -\mathbf{Z}_1^T & \mathbf{Z}_{12}^T \\ 0 & \mathbf{Z}_{21}^T & -\mathbf{Z}_2^T - \mathbf{Z}_2^C \end{pmatrix} \begin{pmatrix} \mathbf{p}_1 \\ \mathbf{v}_1 \\ \mathbf{v}_2 \end{pmatrix} = \begin{pmatrix} \mathbf{y}^F F \\ 0 \\ 0 \end{pmatrix}. \tag{7.11}$$

Finally, the sound pressure level (SPL) at the receiver position is

$$p = \left(\mathbf{z}^C \right)^T \mathbf{v}_2, \tag{7.12}$$

where $\left(\mathbf{z}^C \right)^T$ is the transpose of the vector of the acoustic transfer functions between the patches on the interface 2 and the receiving microphone position in the cavity.

7.4 Experimental Characterisation of Porous Materials

To directly obtain the patch impedance matrix Z of the porous material we have to excite with a surface velocity $\bar{v}_j$ at a certain patch j and block all other patches so that $\bar{v}_{k \neq j} = 0$. Then the impedance relation between patch i and the patch j is given by the ratio of averaged blocked surface pressure to the averaged surface velocity, see Eq. (7.3).

The porous material is a rectangular layer of Basotect® TG melamine foam. In order to block the velocity at the interface steel plates of dimensions 0.2 m × 0.2 m and a thickness of 30 mm have been laid down on the sample. The weight of each steel plate is 9.5 kg. Due to the high mass and stiffness of the plate the rigid body modes of such resulting system appear at very low frequencies, while the first structural modes of the steel plate occur at very high frequencies. In this particular case, all 6 mass-spring resonances occur at frequencies below 40 Hz and the first structural resonance arises at above 2.4 kHz.

The high impedance mismatch between the foam specimen and steel plates, combined with the broad, effective frequency range without resonances allows for establishing proper blocking conditions between 100 Hz and 1.5 kHz. As a consequence, the full frequency range of interest (50 Hz−1 kHz) can be experimentally

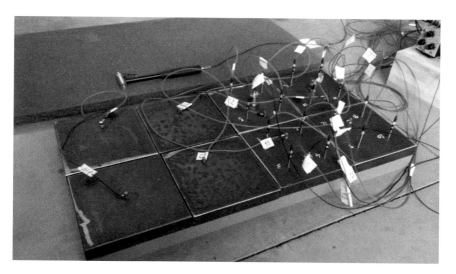

Fig. 7.4 Set-up for the experimental characterisation of a porous trim

investigated. A patch size of 0.2 m × 0.2 m has been therefore chosen for the experimental characterisation. An up-scaling to a larger size which is a multiple of 0.2 m can be easily performed in the post-processing phase by means of the superposition principle. The results that will be shown in the following section are referred to 0.4 m square patches.

Since there are no modes within the frequency range investigated a uniform velocity excitation over the patch can be obtained by hammer excitation in the centre of a steel plate. The imposed velocity on the excited patch has been measured by an accelerometer attached close by the centre of the steel plate. The plates must not be in contact with each other and at the same time the gap between the plates shall be kept as small as possible in order to avoid energy leaks through the slits. The interface pressure has been measured by four ¼″ pressure microphones embedded into each steel plate flush to the foam interface. Note that by doing so only the pressure contribution due to the interstitial fluid is accounted for, whereas the pressure contribution of the skeleton is neglected.

The experimental characterisation of the porous material has been carried out on a melamine sample of dimensions 0.8 m × 0.4 m and the thickness of 42 mm. The material probe has been fully blocked by the steel plates on the top and by a rigid reinforced concrete floor from the bottom. The setup is shown in Fig. 7.4.

7.4.1 Numerical Validation

The experimental methodology for the characterisation of porous materials has been validated by means of an equivalent fluid model implemented into framework of an

FE model. In case that the skeleton motion can be neglected the porous material can be replaced by an equivalent fluid domain with a complex density and a complex bulk modulus. The assessment of these complex quantities can be achieved by means of empirical models such as the one proposed by Delany and Bazley [10]. If the porous material has a high porosity the complex wavenumber k and the characteristic impedance Z_c can be calculated from the flow resistivity σ as a function of the frequency f and the fluid properties ρ_0 and c_0

$$
\begin{aligned}
Z_c &= \rho_0 c_0 \left[1 + 0.057 \left(\frac{\rho_0 f}{\sigma} \right)^{-0.754} - j0.087 \left(\frac{\rho_0 f}{\sigma} \right)^{-0.732} \right] \\
k &= \frac{2\pi f}{c_0} \left[1 + 0.0978 \left(\frac{\rho_0 f}{\sigma} \right)^{-0.700} - j0.189 \left(\frac{\rho_0 f}{\sigma} \right)^{-0.595} \right]
\end{aligned}
\tag{7.13}
$$

with the following validity range

$$
0.01 \leq \frac{\rho_0 f}{\sigma} \leq 1. \tag{7.14}
$$

The validation has been carried out by a numerical model, which mimics the experimental procedure. The FE model of the melamine foam consists of an equivalent fluid with rigid boundary conditions applied on the bottom surface. An imposed velocity has been applied on the excited patch, while the other patches have been blocked. In order to assess the impedance matrix of the full sample this procedure must be repeated for all the patches. The response over each patch has been obtained by integrating the pressure over the patch surface. The input parameters for the numerical model are the speed of sound 343 m/s, the density of air 1.2 kg/m^3 and the foam flow resistivity of 11350 Ns/m^4. These parameters have been measured on a sample originating from the same batch of a material as used during the presented work. For the given flow resistivity the Delany and Bazley model is valid from 95 Hz to 9.5 kHz.

In the calculation the vertical edges of the equivalent fluid have been blocked. A second model has been realised in order to investigate the effects of the boundary conditions applied on the vertical edges of the specimen. In this case the vertical edges are radiating into a defined air volume. The concept of Perfectly Matched Layers [5] is used here in order to avoid spurious reflections at the outer boundary of the acoustic, computation domain.

Figure 7.5 compares the results of experimental characterisation with numerical validation, where the impedance level ($L = 10 \log Z$) and phase are plotted as a function of the frequency. Although strictly speaking the Delany and Bazley model was developed for fibre based materials, an excellent agreement can be observed starting from 200 Hz onwards. The low frequency deviations should not be ascribed to the simplifications of Delany and Bazley model (the Miki correction, for instance, did not influence the low frequency signature), but they turned out to depend on the boundary conditions at the vertical edges, as it will be discussed below.

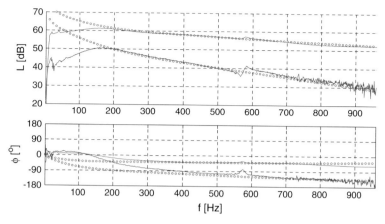

Fig. 7.5 Direct characterisation method of melamine foam with 0.2 m patches: input patch impedance (*red*), transfer patch impedance (*blue*), experiment (*solid lines*), numerical (o *markers*)

The boundary conditions imposed at the vertical edges of the sample strongly influence the patch impedance below 200 Hz. We can distinguish between two limit cases: blocked edges and free edges. These two different conditions have been investigated by the numerical models. The outcome is shown in Fig. 7.6. When the vertical edges are blocked the impedance amplitude is inversely proportional to the frequency and the phase starts at −90 degrees. This trend can be interpreted as stiffness behaviour. In the case the edges are free the impedance amplitude is directly proportional to the frequency in the low frequency range and the phase has a value of +90 degrees. This behaviour represents a mass governed edge radiation impedance. If the patches are defined far away from the sides of the specimen, the

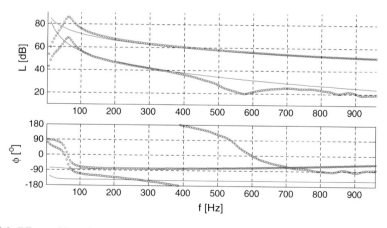

Fig. 7.6 Effects of boundary condition imposed at the vertical edges of the sample on the foam patch impedance; input patch impedance (*red*), transfer patch impedance (*blue*). (i) blocked edges (*solid lines*); (ii) free edges (o *markers*)

edges of the patches are radiating into foam and a continuity type of boundary condition is imposed at the edges of the patch. The boundary conditions imposed on the vertical edges of the trim during the experimental characterisation are however different from the three cases described above, as it was difficult to reproduce well-defined boundary conditions precisely (blocked edges).

Since the effects of the boundary conditions imposed at the vertical edges of the foam strongly influence its low frequency behaviour, it is important to conduct the sound package characterisation with the same boundary conditions as they will occur on the assembled system.

7.5 Validation Study

The PTF methodology proposed has been validated by means of a dedicated test setup consisting of a rigid cavity backed by a dynamic plate, see Fig. 7.7. The rigid cavity consists of reinforced concrete walls coated with hard epoxy paint and has the main dimensions of $1.7 \times 1 \times 0.8$ m. The dynamic plate is made of 2 mm thick steel with all boundary segments being clamped. The inner side of a plate is treated with a layer of porous material, which is the Basotect® TG melamine foam having a thickness of 42 mm in this particular case.

First, the dynamic behaviour of individual, physical sub-domains (plate, trim and cavity) has been determined by component characterisation. Depending on the sub-domain, the impedance matrices and vectors needed for the assembly of the equation system (11) have been retrieved in experimental, numerical or analytical manner. For the plate and cavity characterisation procedures presented in [20] and [21] have been adopted.

7.5.1 Porous Part

As shown in Fig. 7.5, the difference between the amplitude of the input and the transfer impedance is larger than 10 dB starting from some 200 Hz onwards in both

Fig. 7.7 The rigid cavity-backed plate with porous treatment

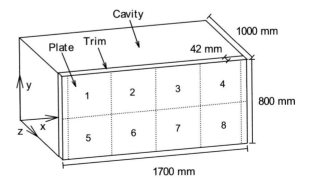

the measurement and simulation. Hence it is possible to consider the material as locally reacting so that the contribution from the patches located far away from the excited patch can be neglected. In the other words, since the material damping is considerably high, only the next neighbouring patch is important for the PTF reconstruction. As a consequence, the impedance matrix of the trim can be approximated by a banded matrix. Therefore, the characterisation of the specimen can be carried out on a small sample just as large as two patches. In order to perform the reconstruction we assumed that the cross impedance of the trim (through the thickness) is equal to the surface one. Since the wavelength inside the interstitial fluid is much longer than the thickness of the foam and therefore wave propagation only exists in the transverse direction, this hypothesis is acceptable in the investigated frequency range.

7.5.2   Assembled System

In order to validate the PTF results based on sub-domain characterisation, full-system measurement campaigns have been conducted on the experimental setup, see Fig. 7.7. The system has been excited by a point force located at position $x = 0.35$ m, $y = 0.5$ m, $z = 1$ m by means of an impact hammer. The sound pressure level (SPL) has been measured in three positions inside the rigid cavity, see Table 7.1 and Fig. 7.8.

Table 7.1 Position of the receiver microphones inside the cavity

	x (m)	y (m)	z (m)
Mic 1	0.270	0.525	0.495
Mic 2	0.870	0.365	0.425
Mic 3	1.460	0.285	0.680

Fig. 7.8 Acoustic cavity with a steel plate partially treated by melamine foam; note the three receiver pressure microphones

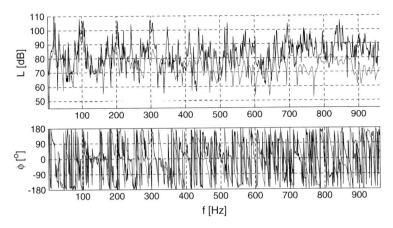

Fig. 7.9 SPL inside the cavity at position 1: reference measurement on the bare plate-cavity system (*black*) and on the trimmed plate-cavity system (*red*)

To estimate the effect of damping induced by the porous layer, a two-step procedure has been adopted. In the first step the dynamic plate has been directly coupled to an acoustic cavity, whereas the melamine foam has been introduced in the system in the latter step.

Figure 7.9 shows the SPL spectra and phases of the reference measurements conducted on the undamped and damped system. The comparison reveals that the melamine foam introduces some damping in the system, but it mainly influences the cavity absorption above 200 Hz due to its rather low thickness of 42 mm.

Furthermore, due to the absence of the heavy layer the trim does not introduce structural damping on the plate. This is can be understood by comparing the patch mobilities of the subsystems, see Fig. 7.10. The input mobilities of the cavity are in

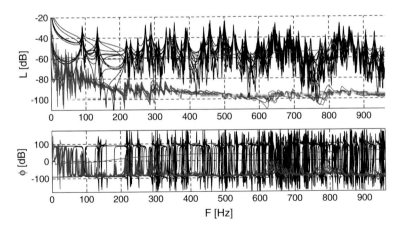

Fig. 7.10 Comparison between the patch input mobility level of the plate (*blue*), trim (*red*) and cavity (*black*)

the same order of magnitude of the trim mobility, while the structural mobilities of the plate are much lower. Only in a limited region below 100 Hz the plate mobility reaches the trim mobility. Thus the main interaction occurs between the liner and the cavity. In addition, the effect of uncertainty of the boundary conditions occurring during the characterisation procedure mainly influences the trim impedance in the low frequency range, where there is just a little interaction with both the cavity and plate. Thus, the effect of the boundary conditions during the characterisation does not have a significant impact on the PTF reconstruction, for this particular type of trim.

Figure 7.11 shows the comparison of PTF reconstruction with full-system measurement of the undamped system. The narrow band prediction of the SPL at position 1 is excellent up to some 430 Hz, which is the theoretical limit for the given discretisation. Nevertheless, the proposed method still yields reliable prediction even at higher frequencies, except for a narrow band between 430 and 500 Hz and around 900 Hz.

Figure 7.12 compares the PTF reconstruction with full-system measurement of the system, where the damping has been introduced via the trim. Here both the plate and trim are characterised in experimentally. The PTH yields an adequate prediction of the full-system behaviour up to the limit of the methodology. As expected the PTF is not able to predict the SPL at the receiver once the frequency is higher than 430 Hz. However the overall trend is captured fairly well.

Figure 7.13 shows again the comparison between the reconstruction and the reference measurement for the damped case, however with trim characterised in a manner this time. This further underlines the high degree of modularity of the PTF process—each sub-system can be characterised in a standalone way by means of the most appropriate technique.

In order to verify the assumption of locally reacting trim, the PTF reconstruction has been conducted with two different, numerically characterised sound packages.

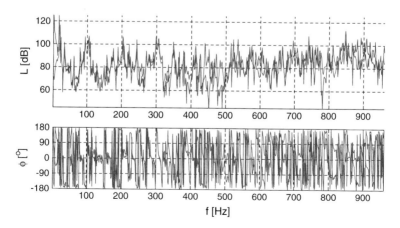

Fig. 7.11 Undamped system—SPL inside the cavity at position 1: PTF reconstruction (*blue*) and reference measurement (*red*). Experimental bare plate and analytical cavity

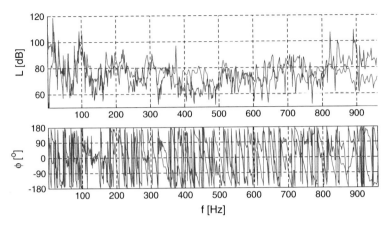

Fig. 7.12 Damped system—SPL inside the cavity at position 1: PTF reconstruction (*blue*) and reference measurement (*red*). Experimental plate and experimental trim

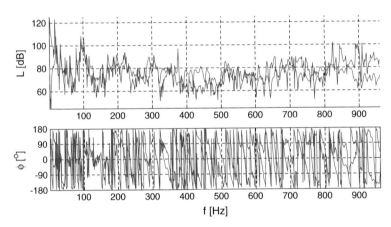

Fig. 7.13 Damped system—SPL inside the cavity at position 1: PTF reconstruction (*blue*) and reference measurement (*red*). Experimental plate and numerical trim

In order to mimic the experimental conditions, first the reconstruction has been performed with a trim having exactly the size of two patches. Next, the PTF reconstruction has been done with a trim specimen having a size of the steel plate (8 patches) in order to account for all transfer and cross (through the thickness) impedances. The results of this investigation are shown in Fig. 7.14.

Apart from the minor differences observed at higher frequencies the two PTF reconstructions do match very well up to 500 Hz. In spite of these small discrepancies we can conclude that below the frequency limit imposed by the discretisation the trim applied can be considered as locally reacting one. Hence, the full trim impedance matrix can be reduced to the input and the next neighbour transfer

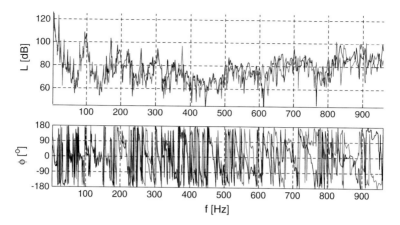

Fig. 7.14 Comparison between the predicted SPL inside the cavity at position 1: PTF reconstruction where the trim is: numerically characterised (*red*) and numerically characterised by modelling a sample of same dimension as of the plate (*black*)

surface impedances. Moreover, this investigation has also proven that the cross impedances can be approximated by the surface ones for this particular porous material.

As explained in [22] the required patch size for valid PTF results is governed by the acoustic wavelength resulting in rather coarse spatial discretisation. This allows for efficient experimental characterisation. The patch averaging scheme for the structural mobility matrix, on the other hand, in principle requires a spatial sampling criterion which depends on the structural wavelength, but the utilisation of multiple sensors on the receiving patch and random hammering on the excited patch reduces the measurement efforts to an acceptable level.

7.6   Conclusions

The PTF approach has been used to solve a coupled vibro-porous-acoustic problem. It allows for an independent characterisation of individual sub-systems and for coupling them at their common interfaces via patch impedance or patch mobility relations. The procedure therefore offers high degree of modularity, meaning that different procedures (numerical, experimental, analytical solution) can be applied based on their merits. In this way, each sub-domain can be characterised by the most efficient technique. Since the sub-systems are characterised in an independent manner, the PTF offers an efficient way to conduct variant studies. In this case only sub-systems subjected to alteration need to be re-characterised.

A novel methodology for the assessment of the blocked impedance of a porous media has been presented and validated. The direct method is fully compatible with the PTF process, hence the information obtained can be seamlessly integrated into the PTF reconstruction of the coupled system.

The measurement methodology described above is however not suited for characterisation of multi-layered materials and it also docs not account for the structure-skeleton interaction. A more advanced experimental material characterisation method, which accounts for these phenomena, is currently under development. Here, the major potential of the methodology proposed consists in the possibility to account for the quasi-global dynamic behaviour of the sound package without the need for a detailed material micro-model.

Acknowledgement The authors acknowledge the financial support of the COMET K2—Competence Centres for Excellent Technologies Programme of the Austrian Federal Ministry for Transport, Innovation and Technology (BMVIT), the Austrian Federal Ministry of Science, Research and Economy (BMWFW), the Austrian Research Promotion Agency (FFG), the Province of Styria and the Styrian Business Promotion Agency (SFG). The research work of Giorgio Veronesi has been funded by the European Commission within the ITN Marie Curie Action project GRESIMO under the 7th Framework Programme (EC grant agreement no. 290050). Finally, the authors gratefully acknowledge the support of COST action TU1105.

References

1. Albert C, Veronesi G, Nijman E, Rejlek, J. (2015) Patch impedance coupling of a structure-liner-fluid system based on direct experimental subsystem characterisation, preprint submitted for review in Appl Acoust
2. Alimonti L, Atalla N, Berry A, Sgard F (2014) Assessment of a hybrid finite element-transfer matrix model for flat structures with homogeneous acoustic treatments. J Acoust Soc Am 135 (5):2694–2705. doi:10.1121/1.4871355
3. Allard JF, Atalla N (2009) Propagation of sound in porous media. Wiley, New York, ISBN 978-0-470-746615-0. doi:10.1002/9780470747339
4. Aucejo M, Maxit L, Totaro N, Guyader J-L (2010) Convergence acceleration using the residual shape technique when solving structure–acoustic coupling with the patch transfer functions method. Comput Struct 88(11–12):728–736. doi:10.1016/j.compstruc.2010.02.010
5. Berenger J-P, (1996) Three-dimensional Perfectly matched layer for the absorption of electromagnetic waves. J Comput Phys 127(2):363–379. ISSN 0021-9991. doi:10.1006/jcph.1996.0181
6. Bertolini C, Falk T (2013) On some important practical aspects related to the measurement of the diffuse field absorption coefficient in small reverberation rooms. SAE technical paper-01-1972. doi:10.4271/2013-01-1972
7. Biot MA (1956) Theory of the propagation of elastic waves in a fluid-saturated porous solid. I. Low frequency range. II. Higher frequency range. J Acoust Soc Am 28(2):168–191
8. Bobrovnitskii YI (2001) A theorem on the representation of the field of forced vibrations of a composite elastic system. Acoust Phys 47(5):507–510. doi:10.1134/1.1403536
9. Chazot J, Guyader J (2007) Prediction of transmission loss of double panels with a patch-mobility method. J Acoust Soc Am 121:267–278
10. Delany M, Bazley E (1970) Acoustical properties of fibrous absorbent materials. Appl Acoust 3(2):105–116. doi:10.1016/0003-682x(70)90031-9
11. Guyader JL, Cacciolati C, Chazot D, (2010)Transmission loss prediction of double panels filled with porous materials and mechanical stiffeners, In: Proceedings of ICA 2010

12. ISO (1996) ISO 10534–1:1996, acoustics—determination of sound absorption coefficient and impedance in impedance tube—part 1: method using standing wave ratio
13. ISO (2003) ISO 354:2003, acoustics—measurement of sound absorption in a reverberation room, 2nd edn
14. Kropp A, Heiserer D (2003) Efficient broadband vibro-acoustic analysis of passenger car bodies using an FE-based component mode synthesis approach. J Comp Acoust 11(02):139–157. doi:10.1142/s0218396x03001870
15. Lanoye R, Vermeir G, Lauriks W, Kruse R, Mellert V (2006) Measuring the free field acoustic impedance and absorption coefficient of sound absorbing materials with a combined particle velocity-pressure sensor. J Acoust Soc Am 119(5):2826. doi:10.1121/1.2188821
16. Maxit L, Aucejo M, Guyader J, (2012) Improving the patch transfer function approach for fluid-structure modeling in heavy fluid. J Vib Acoust 134:051011-1, doi:10.1115/1.4005838
17. Oberst H (1952) Über die dämpfung der biegeschwingungen dünner bleche durch fest haftende beläge. J Acustica 2(4):181–194
18. Ouisse M, Maxit L, Cacciolati C, Guyader J-L (2005) Patch transfer functions as a tool to couple linear acoustic problems. J Vib Acoust 127(5):458. doi:10.1115/1.2013302
19. Pavic G (2010) Air-borne sound source characterization by patch impedance coupling approach. J Sound Vib 329(23):4907–4921. doi:10.1016/j.jsv.2010.06.003
20. Rejlek J, Veronesi G, Albert C, Nijman E, Bocquillet A (2013) A combined computational-experimental approach for modelling of coupled vibro-acoustic problems. SAE technical paper 2013-01-1997. doi:10.4271/2013-01-1997.
21. Veronesi G, Albert C, Nijman E, Rejlek J, Bocquillet A (2014) Patch transfer function approach for analysis of coupled vibro-acoustic problems involving porous materials. SAE technical paper-01-2092. doi:10.4271/2014-01-2092
22. Veronesi G, Nijman E (2015) On the sampling criterion for structural radiation in fluid, preprint submitted for review in J Sound Vib